This book is a result of research that I have done and/or products that I am personally using. Product names and/or Company names associated with those products are as I found them. Any resemblance to actual persons, living or dead, or to actual events or locales is entirely coincidental.

Cover Credit: wk1003mike/Shutterstock

TechBytes
Press
http://www.techbytespress.con

AF809067

Version 2016.09.25

Disclaimer

Unfortunately in this world of lawsuits, I want to

make the following disclaimer. Everything that I have written about here is about my own experiences or as a result of research that I did. Your results may vary from mine. What you will read about here are steps I have taken for my own emergency preparations. If I haven't ordered something but found that it was something that I would be looking at in the future, I will indicate that.

Kindle ISBN - 13:978-0-9971133-0-3
Paperback ISBN - 13:978-0-9971133-2-7

Disclosure of Material Connection: Some or all of the links in this book are "affiliate links." This means if you click on the link and purchase the item, I will receive an affiliate commission. This will not affect the price you pay for the item you purchased. Regardless, I only recommend products or services I use or have used personally and believe will add value to my readers. I am disclosing this in accordance with the Federal Trade Commission's 16 CFR, Part 255: "Guides Concerning the Use of Endorsements and Testimonials in Advertising."

Table of Contents

Your Free Gift

As a special Thank You for purchasing this eBook, when you register your purchase at this URL http://protectyourlogin.com/thac, In return for doing that, I will send you a checklist to help you walk through the process of finding the right tools to protect all of your online accounts and to get the process started of strengthening all the passwords you use. I will also email you when I have updates out to this book so that you can keep the latest copy on your Kindle device or app that you are reading it with.

There are other benefits you get from registering your copy of "The Hackers Are Coming - How-To Safely Surf The Internet". Once you have registered the book, you will receive a special URL where you can get access to a special set of worksheets that I have designed as a companion to this book. Once you have purchased your first set of worksheets, you will have lifetime access so you can always get a fresh set of worksheets at any time. In the coming months, I will also be creating members only site where you can get access to information that has been released since the book came out. There will also be other benefits you will see once the site is online. This, like the worksheet, is only available to those that have registered their purchase of this book.

Thank you for purchasing this book and for taking the time to read it.

Introduction

I will be the first to admit that I should have been using better password sooner than I did. What drove this home was when the some of the accounts of two individuals that I know were compromised. While it didn't take a long time to get control of their accounts back, not everyone is so lucky. In one of the cases that I am familiar with, there was also a loss of income because the operations of one of the businesses had to be shut down for more than just a day or two.

Given enough time and resources, if someone wants to get into one of your accounts, they can. The key is to make it hard enough for them to get them discouraged to where they will move onto other low hanging fruit or easier to hack targets.

Hardening your accounts is not something you will do overnight. Where possible, I would recommend that you do it in stages. First make your passwords stronger and get used to working with them. While you are going through the process of increasing the level of protection on your accounts, make note of which ones support Two Factor Authentication or other methods where you are prompted for a second piece of information to verify that you are the actual account holder trying to login.

If you see any accounts not listed and would like to see what options are available, please reach out to me via my website - http://protectyourlogin.com and I will see what I can come up with for you. That will be added to this book and you will be able to get an updated copy of this book containing it.

About The Author

Ron Nutter has written for technical trade press since 1990. Over the years he has written for InfoWorld, LAN Times, TechRepublic, NetWare Solutions, Network Solutions and Network World. He has written and produced HelpDesk ToolChest with Ron Nutter for Network World.

The list of people interviewed includes Steve Gibson of GRC Research, Laura Chappell of Chappell University, Chief Twit Leo LaPorte and Dave Ramsey of Financial Peace University to name just a few.

Over the years, Ron has obtained a host of certifications -

Cisco - CCDA, CCDP, CCNA, CCNA Security, CCNA Wireless, CCNP Security, CCNP Route/ Switch

Microsoft - MCP, MCSE

Novell - CNA, CNE, Groupwise CNE, ECNE, MCNE

VMware - VCP5

If you would like to see about reviewing your product for the website, video podcast, audio podcast, or possible inclusion in a book, please contact me at helpdesk@networkref.com. When shipping items for review, please use only Fedex or UPS and send the tracking number so that arrangements can be made for someone to be at the office when delivery is attempted.

Ron is also available for a limited number of speaking and consulting engagements.

Register your book and get notified of updates - http://www.protectyourlogin.com/thac.

Bookmark this link to keep up with new books as they are released - http://www.ronnutter.com/booklist/.

Terminology

There are several terms that we need to define before you get very far into this eBook.

2FA - short for Two Factor Authentication

Application Password - this is a special, unique and long password used by an application to login to another service. This is used when the application in question doesn't understand or have the ability to use two factor authentication to login to the service it is trying to access.

FIDO - Fast IDentity Online (FIDO) Alliance. Organization that created the U2F and UAF standard(s).

Hardware Token - this can take on a variety of forms. The more sophisticated hardware tokens contain a LCD display that displays a set of numbers that change every 30 seconds or at an interval defined by the manufacturer. Other hardware tokens plug into the USB port of your desktop or laptop.

OTP - One Time Password, a password that can only be used once and then it expires

SMS - acronym for text messaging service on your smartphone

Software Token - this is an app that runs on your smartphone or computer that generates a periodic set of numbers used during the login process

Two Factor Authentication - uses your login name and password (something you know) and your two factor token password (something you have) to authenticate yourself to the site/service you are trying to login to

UAF - Universal Authentication Framework. This is password-less experience where something like a fingerprint is used for the password. Can be looked as a replacement for simple authentication.

U2F - open authentication standard that strengthens and simplifies two-factor authentication. Possibly a replacement for today's time-based, second factor authentication.

QR Codes in this book

You will see a series of QR codes in this book. Whether you are reading this on a laptop or a kindle reader, these codes will give you a way to go to the information that the QR code is for. Since this is something that you may not have worked with before, I will give you the links to the resources I have used. This is a new way of making books interactive and I hope it will increase the usefulness of this book to you.

The first QR code will be for your smartphone -

Android - http://www.tinnitusmystory.com/aqrc

IOS - http://www.tinnitusmystory.com/iqrc

Both of these apps are from the folks at Scan, Inc. I like use the same app from the same vendor on any platform I am working on. Found that it helps minimize troubleshooting issues if they arise.

This next series of apps will be for laptops and desktops. My criteria for selecting an app for this platform was that the potential for not having a camera was a possibility. My reasoning is that if I am reading a Kindle book on my laptop/desktop that having a camera to read the QR code from the book wouldn't work since I couldn't point the camera that is built into the screen in a direction that it could read the screen. I went with apps that would essentially be a screen scraper like app that would allow me to read the code and be able to take action with the code.

Mac - http://www.tinnitusmystory.com/mqrc ($3.99)

Windows - http://www.tinnitusmystory.com/wqrc

Here is a test QR code to work with. It should send you to my website - http://www.ronnutter.com.

Some the QR codes or URL links in this book may be affiliate links that I will get a commission on if you buy the item. This will not increase the cost of the item you are purchasing.

How to use this book

There is a method to the madness in this book. Please take a few minutes to read this and everything else will make sense. The first thing that you will go over are all the tools that are at your disposal to begin the journey to putting the most protection in place that you want for all of your social media accounts.

For those that remember the catalogs of years gone by that had items organized in categories of good, better and best. That is how the steps I outline for each of the sites I mention here are organized. The first level, putting a strong password place puts you at the good level for protecting your accounts. I wrote each of the procedures in the same form under the presumption that you may have some but not all of the accounts that I written up. I assume in each of the steps that you have already selected a password manager app to accompany you on this journey.

Using a single strong password is better than using something like Password or Cisco123 for your account password. Having a unique, strong password is even better. Having a distinctly different password for each account increases the level of difficulty for the attacker by adding another barrier to entry in compromising more than one account at the same time. Taking your preparations to that level gets you to the Better level of your journey. Creating a strong password is something that may not come easily to some. This is where the right password manager application or even some independent tools built just to help you create a strong password according to criteria you specify in terms of length, use of numbers, and special characters (i.e. punctuation) will help you put together a difficult password to guess for an attacker.

Using Two Factor Authentication, using either a token that has a constantly changing set of numbers on a software or hardware token or a SMS text message with an access code to use, will help you establish a barrier to your accounts that should help put as much protection in place for your accounts as is possible. Don't let this type of protection scare you because of the length of the name. Once you have enabled this on one or two of your accounts and get used to the slightly different login process, you will be ready to pursue that same level of protection for your accounts.

More websites are starting to implement a process where you have to select and provide answers for a series of questions. These questions may be used as an additional step to go through for logging into an account or as a protection step before allowing an account password to be changed. This is another mechanism to have in place for those websites that use it. I will talk more about this in a later section of the book.

Each service type listed in the book is written so that it assumes you don't have any other accounts to register. It is meant to be a checklist to make sure you go through each of the steps needed to better protect your account.

Before you get started reading this book, keep this next fact in the back of your mind. According to the folks at IDKey (a company you will see mentioned later in this book), it only takes a hacker 10 minutes to guess a 6 character lower case password. That alone should be a good reason to increase the difficulty or strength of the passwords that you currently use.

When you see the word **VIDEO** in a chapter, the URL and QR code that follows directly link to a video on YouTube. This goes into more detail about the chapter you are reading or about the product or technology just discussed. If you want to skip ahead, all of the Videos for this book are in the same playlist on my channel, TechBytesRN, on YouTube. Your first video for this book is below. The other videos will be in the book similar to what you now see.

Watch Video -

http://protectyourlogin.com/trailer

What is the FIDO Alliance ?

I normally don't get that excited about a standards group. The FIDO (short for Fast Identity Online) Alliance is a little different. The group formally launched in February 2013 by six companies to lead the industry by moving from password dependencies to strong authentication which is more secure, private and easier to use than passwords. Of the companies that I provide how to use Two Factor Authentication with in this book, it should be noted that Paypal was one of the founding members of the alliance and continues to remain involved.

Who would think that a year later, you would see the membership swell by to more than 100 companies including Google, Microsoft, and Netflix to mention just a few of the members. The number continue to grow, sometimes daily. When I checked the current membership list in October 2015, it was well over 225 members and shows no signs of stopping. Take a look for yourself to see the level of involvement and who is involved - https://fidoalliance.org/membership/members/. When this many companies are involved, we all win with this level of cooperation.

One of the first ways that I heard about FIDO was when I initially heard about one of the members of the FIDO Alliance, Yubico. I had a long conversation with the CEO and Founder of Yubico, Stina Ehrensvärd. Our conversation lasted for almost 45 minutes and was the best education I have ever received on a technology that was new to me.

The adoption of FIDO is continually growing. The FIDO 1.0 Standard was released on December 9, 2014. Since that time, Microsoft has announced support in Windows 10. NTT DOCOMO, deployed FIDO authentication for its subscribers in Japan, becoming the first Mobile Network Operator to deploy that technology. Dropbox, a well know cloud storage company, recently announced it's support for FIDO and has made it available to its users. Github has announced that it will accept FIDO U2F authentication on its popular code-collaboration service.

Most of the ways I show you in this book on how to protect your online presence are provided by FIDO members (Yubico, for example). As I find other FIDO certified products that are a good fit, I will update this book as those solutions become available. Here is a YouTube video that I would encourage you to take a look at to get a good understanding of what FIDO is and how it can help protect you. Here is a page that will have the latest videos that show other uses of FIDO certified products - https://fidoalliance.org/adoption/videos.

CHAPTER ONE

Tools

What tools are the best to use ?

What I will outline in this part of the book will be the tools I have looked at and from those selected the ones that worked best for my situation. There is no one solution that will work for everyone, so I wanted to provide you a list of what I had looked at to save you some time when you are looking for the combination of tools that will work for you.

Your first stop is to select a Password Manager. I had tried several open source ones when I first started using one several years ago. The ones that I had looked at were a little clumsy in the area of file synchronization area. When I was able to work around that, dealing with delta changes was a little more challenging. Delta changes are when you make change A on one device and change B on a different device and getting them to apply in the correct order when synchronizing all devices so that the changes where applied in the right order.

You will read about Two Factor Authentication. Don't let the term scare you. It is the next step beyond username and passwords to access a site/ service. Using a strong password will only protect you so long. It is just a matter of time and resources before someone with the right tools and determination before a password can be cracked. Using Two Factor Authentication increases the level of difficulty to the point to where only the most determined hacker may try to stick it out. Since they have bills to pay as well, they will only spend so much time without having any financial gain.

When using Two Factor Authentication, you will need to look at using either a Software or Hardware Token. I have used both and they each have their own pluses and minuses. Having dealt with the costs of a hardware token from one of the well known commercial Two Factor Solution providers, I initially shied away from that option. That was until I had the chance to talk to Stina Ehrensvärd, CEO and Founder of Yubico. Using the Google Authenticator client was as safe as I thought but once someone had my device that the Authenticator was installed on, they essentially had the keys to the kingdom for accessing all of my Two Factor enabled accounts. That is what got me to start looking the Yubikey solution. Once I got a Android smartphone containing NFC, I was able to experience the full Yubikey solution. I get the advantage of a Two Factor software token with the security of a Hardware Token.

If a site doesn't support fully support Two Factor Authentication but offers you a SMS or Email with an access code, that is better than nothing. The challenge with using SMS is that depending on what provider is being used to send the SMS message to you could result in a delay of several minutes before you get the code needed to complete the authentication process.

Crafting a strong password is not something that comes easy to everyone. That is complicated by some websites not allowing you to use special characters that some of us would know as punctuation. There are a variety of tools available, some available in the password managers that I talk about in this book, to help make that process a little easier. One rule that I am doing my best to not use the same password on multiple sites.

One recent account hack that I learned about was using answers to security questions that some sites use as an additional level of protection to make sure that only the actual account holder is requesting a password change or as another step in the login process. The problem is that in the case I am familiar with, the answers to the security questions could be readily found in information available online about the account holder. Depending on how public of a persona that you already have, or may have at some point in the future, giving the correct answer could make it one step easier for someone to gain access to your accounts. The two ways that you can answer the security questions and help avoid this will be to give either inaccurate answers or give an answer that is a series of letter, numbers and functional or special characters. In either case, be sure to record both the question and the answers you give for that sites entry in the password manager that you decide to use.

Password Managers

mSecure
http://protectyourlogin.com/msec

This is the password manager that I have used for several years. I had used a variety of open source password management apps but the problem I ran into was that they were only for the PC or had a file that had to be manually copied between machines to keep the passwords in sync between systems. That is what attracted me to the mSecure software family. It supports just about every platform that you would probably run into on a daily basis.

You have two different options for keeping your logins and passwords in sync. The first involves having all the devices on the same wifi network. That is sometimes easier said than done. What I have implemented is syncing the password file to my Dropbox account. In that way, whenever I bring up mSecure on whatever platform I choose, that I automatically have the latest password file available for use. The real silver lining in this is that if I have made multiple changes on the same or different logins, mSecure applies the changes in the chronological order they were made in.

LastPass

http://protectyourlogin.com/LP

LastPass is a cross platform password manager. To really get full functionality, you will need to subscribe at the Premium level (as of September 2015, that is $12 per year). For that amount, you get the ability to have unlimited sync across an unlimited number of devices. You also get the ability to support more MultiFactor Authentication systems such as Yubico and biometric options. Priority support come with this package. If you operate this on a locked down computer, you will also have the option LastPass from a USB drive when at the Premium level.

keeper

http://protectyourlogin.com/Keeper

Keeper Security is a multi platform password management and security tool. Pricing has two different levels - $9.95 give you password sync with one device. You also have the ability to add additional cloud file storage. For $29.99 a year, you get the ability to sync multiple devices.

dashlane

http://protectyourlogin.com/DL

There are two price levels - free and $39.95. The main distinction in one device on the free plan and multiple devices for $39.95. All platforms except for Linux are supported.

1Password

http://protectyourlogin.com/1pass

All platforms except for Linux are supported. The Windows and Mac clients are fully functional for 30 days. The licenses for Mac and Windows start at $49.99. There is a charge for upgrade that varies depending on the situation.

SplashID Safe

http://protectyourlogin.com/splash

There are two versions - free and two different levels of pay (monthly or yearly). Pro users get automated backups. The last 5 backups can be downloaded and restored. All platforms are support except for Linux. By implementing the Chrome extension, you can integrate SplashID Safe functionality directly into the browser.

Software Tokens

Google Authenticator

Watch Video -
http://protectyourlogin.com/ga

This is the first two factor client that I got started with. It is easy to use and is something that you will have something up and running in a few minutes. This is intended for a phone that has a camera in it cause you need to capture the QR code that is displayed as a part of the two factor enrollment process.

One suggestion that I found out about that has proven very helpful is to do a screen capture of the QR code used for each login that you will use this, or any of the two factor clients that I will talk about in this eBook. Put that screenshot in an Evernote folder or whatever app you use to keep track of important items. Label that screen shot as to what website it is for and the date you started using it. This is available on both Apple and Android platforms.

SAASPASS

This is the latest two factor client that I have found. Of all of them, it is the closest to being a swiss army knife in that it does a little bit of everything. What separates this one from the others is what it can do beyond just giving you a 6 digit code for login verification. You can add an additional software component that you can either scan a code with the SAASPASS app or click the Remote Unlock button for that computer in the Computer Login Section of the app.

When you add a website login, when possible, SAASPASS will also put an appropriate graphic for the site to help you quickly identify it from the list of the other logins that are already in the list. This is available on both the Apple store and Google Play store.

Authy

This was the first two factor client that I found after I had been using Google's Authenticator client for a while. What distinguishes Authy from Google's client is that it backs up the tokens you have it configured for to its own cloud. In addition to that, it also makes sure that the device that is running the Authy client is on the correct time so that you don't have unpredictable results caused by a different in time references. If you are converting from Google Authenticator, there are instructions on Authy's website to help you make that transition. This is available from the Apple store and the Google Play store.

Symantec VIP

This two factor client isn't as widely used as the other two factor clients I have listed above. I have seen this used mainly for eBay and PayPal. If you go with this 2FA client, there will be a token serial number that you will want to record. You will want to make sure you have your contact and recovery information up to date. During an upgrade to my smartphone, the token serial number changed and I didn't catch it. My first indication of a problem was when I lost access to getting into either of the accounts I had it configured for. It took about 20 minutes on the phone with an understanding customer service again to gain access to my accounts and setup a new security token. This is available for both Apple and Android users.

Yubico Authenticator

Watch Video -
http://protectyourlogin.com/ya

This client is just one part of a three part solution. In order to get this to work, you will also need Yubico's NEO token. You will need a phone that fully supports NFC (Near Field Communication). At the point, this means that only Android phones can be used. While Apple does have NFC support in the iPhone 6 and later, it can only be used with the Apple Pay system. I have successfully used the Yubico Authenticator on my Google Nexus 6 phone This app can be downloaded from the Google Play store.

Hardware Tokens

Yubico U2F

This is my go to hardware token. I became aware of this token when I was writing a previous book. I had been using the soft token client released by Google. After a long talk with the CEO of Yubico, Stina Ehrensvärd, I learned of potential security issues with software only clients. The biggest one was that when using a soft token and the phone that it was used was stolen, you have just given a major component to your account security to someone who will use it for no good. This is when I started using the Yubico key.

I always like to have a backup plan in everything I do. Imagine my surprise when I learned when setting up the U2F Special Security Key that, at least on Google accounts, that I was able to assign multiple keys to the same account. This is handy as I can have one key at home and another key with me and be able to get into the account without having to get the wallet the one key is in. Use the QR code below or click on the

URL to purchase your own key.

There is one item you need to know about. The U2F functionality only works with Google's Chrome Browser at this point. I have been on a conference call where Yubico talked about working with Firefox, IE and Edge. That is work in progress and dependent on third parties so there isn't a definite day when this functionality will be available on other browsers but there is at least a roadmap for this happening.

http://www.ronnutter.com/yubu

Yubico VIP

When you are purchasing an item in an auction you just won on eBay or paying someone using your PayPal account, this is the key that you will want to use. There is a soft token that you saw mentioned in the previous chapter but that can have it's own challenge as you can see. This token is supported by both eBay and PayPal. Due to a security requirement of the Symantec authentication system that both of these sites use, the functionality needed for the key to work can only be implemented at the time of manufacture. Be careful about making any changes to this key as once you have deleted or damaged the functionality needed by eBay/PayPal, you will need to purchase a new key. Even though Yubico has other blank hardware tokens, you will notice a large checkmark on the back of the token. Use the QR Code below or click on the URL to purchase your own key.

http://protectyourlogin.com/yusy

Yubico NEO

This is the latest key in the Yubico family that I have been working with. The graphic you see is actually two different keys. The NEO is the larger of the two keys show. The NEO-N is the smaller key. I have been using the NEO with my Google Nexus 6 with some very interesting results. Since the NEO supports NFC as well as my Google Nexus 6, I can use this key in conjunction with the Yubico Authenticator (available from the Google Play Store) to access all of my accounts that are protected with Two Factor Authentication (with the exception of eBay and PayPal because they use a different system). I bring up the Yubico Authenticator app on my Nexus 6 and tap the Yubico NEO on the back of the phone. It immediately populates the screen with all of the tokens I have programmed it for. Once the currently displayed values expire, I have to tap the NEO on the back of the phone to get updated values displayed. If I share the phone with someone, I can either let the current keys expire (they will turn grey on the screen) or reboot the phone to clear the codes from the screen. Use the QR Code below or click on the URL to purchase your own key.

Yubico has heard requests for supporting Apple users. The problem appears to be with Apples implementation of NFC. They are working on a different key with Bluetooth support that will work with the iPhone. I have requested to be admitted to the beta test program. At this point, I am waiting to receive a device to test with. I will update this book when that has happened.

http://www.ronnutter.com/yune

Yubikey 4

This is the latest addition to the line of Two

Factor Keys from the folks at Yubico. One of the big differences with this key is that you have the option using most of the major browsers instead of only being able to use Chrome. This key has the potential of setting a new standard in functionality since it supports supports multiple authentication protocols, including Yubico OTP, smart card (PIV), and FIDO U2F.

The Yubikey 4 key is supported by the following password manager applications - Keepass, Password Safe, LastPass, Password Tote, pwSafe, and PassPack. Another advantage to this new key is that you have the ability of using it as a source of two factor authentication for Full Disk Encryption. The folks at Yubico have a pdf available for download on their site going through the process of integrating the Yubikey with Full Disc Encryption to get you started on that path when you are ready.

http://protectyourlogin.com/y4

Injector

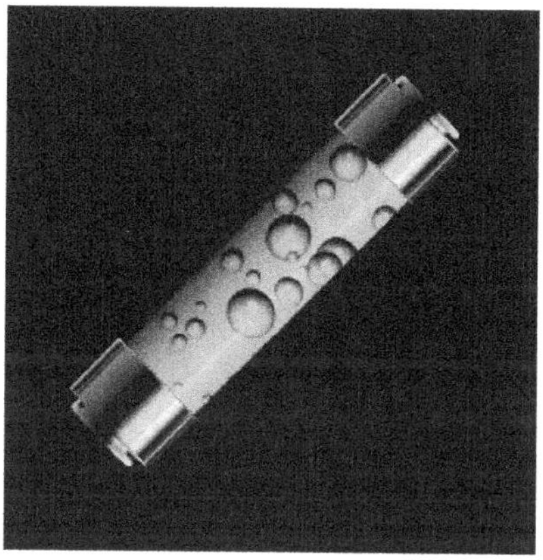

This particular solution is one that is hard to classify. Here is why - it is part hardware and part software. For the app that resides on the phone, it is a password manager but when you pair it up with the matching Bluetooth LE adapter, you have a solution that can automatically send just your password or username and password, depending on how you have it configured to work. When I first found this solution, I tried with two different bluetooth sources seeing if it would work with any bluetooth device and found that it wouldn't. This is good and bad. It would be nice if you could use the existing Bluetooth adapter present in most laptops. On the other hand, by using their adapter, a higher level of encryption can be used between the smartphone/tablet running Injector app and the Injector BLE adapter.

What got my attention was that this was the first solution I had found that had multiple browser support. At this point, it support Safari and Chrome. Firefox is on the roadmap but there was no date I could find on the website to indicate when it would be available. Most of the solutions I have found up to this point currently only support the Chrome browser.

Once I got one of the two Bluetooth LE adapters from the folks at Password Injector, I got the chance to really see the potential of this system. It took a little getting used to from a configuration standpoint but I think you will be pleased with what you see. During the credential setup process for a particular service/site, you will need to scan the QR code for the information needed to generate a OTP (One Time Password) used for sites that allow you to use Google's Authenticator or similar apps. This is where having a screenshot of the QR code when you initially setup the login for two factor authentication will be handy. A nice side benefit of using the Injector app for 2FA (Two Factor Authentication) is that unlike the other apps that I have looked at for this purpose, it doesn't display the code on the screen. The only time you will see the code is when the code is sent to your computer when the Bluetooth adapter is installed.

http://protectyourlogin.com/painj

Watch Video - With 2 Factor
http://protectyourlogin.com/piupt

Watch Video - Username & Password
http://protectyourlogin.com/piup

IDKEY

The IDKey is the first hardware token that I have found that supports use of biometrics (i.e. your fingerprint) as a method of authentication when logging in to a supported service. It supports both Bluetooth and NFC which means that it should be an option for both Android and iOS users. I will be writing more about this device when I receive one and can show you how I have implemented it.

http://protectyourlogin.com/idkey

Email and SMS as Two Factor Tokens

With some websites, getting an email or SMS message with the OTP (One Time Password) to use for this particular login attempt is your only option. While not idea, this is better than nothing. Which will work best for you depends on how much you travel. If you travel out of the country, getting a SMS message may be problematic (and possibly expensive) since getting a SMS message can difficult when you are operating on a different carrier in another county.

If you only have a SMS message as a method for delivering the authentication token, consider getting a Google Voice account. This gives you an option for getting the SMS message via an email. While not ideal, it is an option worth thinking about.

Crafting a strong password

Creating a strong password isn't one of my favorite things to do. There are several ways that you can make the process a little less painful. Someone I worked with several years ago taught me a trick that has served well on several occasions.

Take a password that more than one of us has used, our old friend "password". With a few changes of upper/lower case and punctuation and you can have something that looks like P@s$w0rD. You can make it a little more challenging by adding an exclamation point at the end and even a few number, in no particular order of course. What you want to avoid is having a password that could be guessed using what is known as a "dictionary attack" where a series of passwords is tried using the words right out of a dictionary.

If you want to make it a little more challenge for an unwelcome intruder, most password managers offer the ability to come up with a randomly generated password as a part of the process of entering the information about a particular login that you want to keep track of. I have run across some websites that won't let you use "special characters". That usually seems to mean punctuation letters. The website should just say that instead of making you guess what they don't want to use. In addition to the password manager, there are also several apps available for both Apple and Android phones that can help you come up with a challenging password as well.

Watch Video -
http://protectyourlogin.com/gsp

Account Security Questions

More websites are starting to ask one or more questions that you can answer that can be used as a secondary challenge during the login process. In some cases, these same questions can be used as a part of the process you can go through when you have forgotten the password you setup for the account. Some sites may allow you to choose from a list of questions to have in your profile or even allow you to enter your own questions.

There is something that you need to think about when both selecting an answering the questions. While our basic intent is to answer a question with a response that we can remember, you may want to think a little differently about that. One reason to do so is that some answers you might provide, in this day of being able to find just about anything about yourself on the internet, you need to think about providing an alternate answer that doesn't match what can be found about you. For example, if one of the security question options ask about your place of birth and you were born in New York, pick another city/town such as Orlando or Knoxville in an area of the country where you haven't lived or isn't geographically near where you currently reside or where you were born.

If you feel up for the challenge, you can think about taking it to the next level. Some websites offer a wider variety of questions to choose from. One thing to think about is to not use the question for where you were born on every site where that is an option. Depending on the number of sites up have accounts on, you may not have a large variety of questions to choose from. The point here is to make it challenging enough to encourage someone attempting to access your account to go elsewhere.

Keeping track of both the questions chosen and the answer you gave for each question, should be kept in a note field in the password manager app you use. Depending on application you have chose, you might want to think about setting the note field as a "sensitive" field. In doing so, it helps keep someone from looking over your shoulder and learning some of your answers. Depending on how sophisticated your application is, you could have the question be in plain text in one field and the answer be hidden in a sensitive field right below it. In that way, you would have one more layer of protection about the answers you have used.

What may be easier than using alternate location names is to use a random set of letters and numbers instead of a plain text answer such as the name of a city or brand of vehicle you had. This increases the level of difficulty of someone successfully guessing your answers or finding the answer on a social media site or by the use of a search engine. The importance of either using a false location or random letters for an answer recently hit home with a well known internet marketer. Every time the security questions were used to regain access to an account, the attackers knew about it because they were watching the email account for this internet marketer. After the password and been changed, they had changed it back to something they knew moments later. It took approximately two weeks to figure out how the hackers kept getting access to the accounts to put a system in place like I have outlined here to put a stop to their activities.

In addition to using false or misleading answers for security questions, also implementing a two factor solution from vendors such as Yubico or Password Injector from Bluink Ltd, might have prevent the situation from happening to begin with. While using a strong unique password for each website or service is important, that is just one tool in protecting your accounts from being taken over by someone else. The more steps you take now, increases the level of difficulty for someone trying to break into your accounts.

CHAPTER TWO

Backup & Sync

Dropbox

Use a Strong Password

For those that aren't familiar with this term, a Strong Password is one that contains (where permitted by the website or service) a combination lower case letters, uppercase letters, numbers, punctuation or other special characters to make up a password. The longer the password is, the long it will take for someone to guess or hack your password. This is an area where I am probably as guilty as everyone else is on this. I have a "favorite" password or two that I liked to use in the past. This kept me from having to have a way to keep track of all the passwords that I used on different websites. With more websites adding additional layers of security such as challenge questions, etc., I have had to start using a password manager app to help keep track of the different passwords, the challenge questions and answers used on a particular website, the recovery procedure if I am locked out of a website, what additional login procedures I have used for that site, etc.

Use a Unique Password

This one will be a bit of a pain but the time taken to do this will pay off in the long run. Look at it this way, if one of your logins on a particular site is compromised or hacked, you have taken a step to minimize the potential for additional website logins to be breached as well. Most password manager apps have the functionality of

helping you generate a unique password. There are other apps that do this as well. A search of the app store for your phone or the internet for you laptop/desktop should show you some options to look at.

Different websites have different rules for how you can make your password strong. Some will let you use "special" characters such as * or +, others wont let you use those characters but will make sure that you don't use a password that can be looked up as a dictionary word. Some will not allow the same character or number to be used repeatedly. Remember, the more uniqueness you can have in a password makes it that much more difficult for someone to hack your account.

Enabling Two Factor Authentication

- Login to your account
- Click on your login name
- Click on Settings
- Click on the Security tab
- Click on Enable beside Status on Two Factor Authentication
- Click on Get Started button
- Enter the password for your account

You can select SMS, Two Factor client, or Security Key for the level of Two Factor Authentication you want to enable for your account. The implementation of all three of the following options isn't needed. If you enable SMS and using the Google Authenticator client, you are given the option of using the Google Client first. If you check the option on the screen about not being able to use the software authenticator, you will be given the option of using a code sent via SMS to your phone number on file. As a last resort, you have the option of using the Emergency Backup code.

Using SMS for Two Factor Authentication

- Select Use text messages
- Click on Next
- Enter the number for your Smart Phone
- Click on Next
- Enter the code that is sent to your Smart Phone
- Click on Next
- Click on Next to go past entering a backup phone number to send the code to
- You will see a verification screen verifying enrollment in Two Factor Authentication
- Write down the emergency access code (put this in your password manager app)
- Click Enable two-step verification button

Use Two Factor Client (i.e. Google Authenticator or SAASPASS)

- Select Use a mobile app
- Click on Next
- Do a screen capture of the QR code displayed on the screen
- Put the screen capture in a safe place
- Click on secret key link
- Capture the secret key as an option to login if you don't have your phone

- Past the secret key in your password manager app
- Click on Next
- Enter the 6 digit code displayed by your two factor client
- Click on Next
- Click on Next to go pass the screen for entering a backup phone number
- Click on Next
- Copy the Emergency access code and put in in your password manager app

Use Security Key

The key I used for this part is the Yubico FIDO U2F Security Key

At this point, you should still be in the Security Tab portion of your account.

- Click on Add beside Security Key
- Enter the password for your account
- Click on Next
- Click on Begin Setup
- Insert the Yubico FIDO U2F Security Key
- Click Key Inserted
- Tap on the gold button on the key when prompted on the screen
- When the key is registered, you will get a message on the screen.
- Click on the Finish button

Test Two Factor Authentication

- Logout of your Dropbox account
- Log back in and test the two factor

authentication

- Enter your username and password
- Follow the steps on the screen as indicated for the two factor authentication you have setup

With Dropbox, you have the option of using the Two Factor Authentication option you initially setup before using the FIDO U2F key. If you don't have that key available, click Use mobile authenticator option to use that option instead.

Watch Video -

http://protectyourlogin.com/db

Evernote

Use a Strong Password

For those that aren't familiar with this term, a Strong Password is one that contains (where permitted by the website or service) a combination lower case letters, uppercase letters, numbers, punctuation or other special characters to make up a password. The longer the password is, the long it will take for someone to guess or hack your password. This is an area where I am probably as guilty as everyone else is on this. I have a "favorite" password or two that I liked to use in the past. This kept me from having to have a way to keep track of all the passwords that I used on different websites. With more websites adding additional layers of security such as challenge questions, etc., I have had to start using a password manager app to help keep track of the different passwords, the challenge questions and answers used on a particular website, the recovery procedure if I am locked out of a website, what additional login procedures I have used for that site, etc.

Use a Unique Password

This one will be a bit of a pain but the time taken to do this will pay off in the long run. Look at it this way, if one of your logins on a particular site is compromised or hacked, you have taken a step to minimize the potential for additional website logins to be breached as well. Most password manager apps have the functionality of

helping you generate a unique password. There are other apps that do this as well. A search of the app store for your phone or the internet for you laptop/desktop should show you some options to look at.

Different websites have different rules for how you can make your password strong. Some will let you use "special" characters such as * or +, others wont let you use those characters but will make sure that you don't use a password that can be looked up as a dictionary word. Some will not allow the same character or number to be used repeatedly. Remember, the more uniqueness you can have in a password makes it that much more difficult for someone to hack your account.

Enabling Two Factor Authentication

- Login to your Evernote account
- Click on the Account icon in the lower left hand corner of the screen
- Click on Security Summary under Security
- When the Security Summary screen appears, click on enable beside Two Step Authentication
- Click on Continue
- Click on Continue on Important Things to Remember screen
- Click on Send Verification Email button
- Enter the code received in email into the designated field on the screen
- Click on the Continue button
- Enter your smart phone number where indicated on the screen

- Click on the Continue button
- Enter the code you received in a SMS message where prompted on the screen
- Click on the Continue Button
- Click on Skip to skip past the enter a backup number
- Click on Continue with iOS or Android depending on the smart phone you are using
- Register the QR code as instructed
- Enter the 6 digit code displayed
- Click on the Continue button
- Enter the password for the account and click on the Continue button
- Click on the Verify button

Test Two Factor Authentication

- Logout of your Evernote account
- Log back in and test the two factor authentication
- Enter your username and password
- Follow the steps on the screen as indicated for the two factor authentication you have setup

With Evernote, you have the option of using the Two Factor Authentication option you initially setup before using the FIDO U2F key. If you don't have that key available, click Use mobile authenticator option to use that option instead.

Watch Video -
http://protectyourlogin.com/en2fa

Microsoft One Drive

Microsoft One Drive is a cloud storage drive where you can keep files without having to worry where your flash drive is. It should be protected just like any other online account that you have. By following the steps I have outlined, you will do as much as you can to keep others from trying to get into your account.

Use a Strong Password

For those that aren't familiar with this term, a Strong Password is one that contains (where permitted by the website or service) a combination lower case letters, uppercase letters, numbers, punctuation or other special characters to make up a password. The longer the password is, the long it will take for someone to guess or hack your password. This is an area where I am probably as guilty as everyone else is on this. I have a "favorite" password or two that I liked to use in the past. This kept me from having to have a way to keep track of all the passwords that I used on different websites. With more websites adding additional layers of security such as challenge questions, etc., I have had to start using a password manager app to help keep track of the different passwords, the challenge questions and answers used on a particular website, the recovery procedure if I am locked out of a website, what additional login procedures I have used for that site, etc.

Use a Unique Password

This one will be a bit of a pain but the time taken to do this will pay off in the long run. Look at it this way, if one of your logins on a particular site is compromised or hacked, you have taken a step to minimize the potential for additional website logins to be breached as well. Most password manager apps have the functionality of helping you generate a unique password. There are other apps that do this as well. A search of the app store for your phone or the internet for you laptop/desktop should show you some options to look at.

Different websites have different rules for how you can make your password strong. Some will let you use "special" characters such as * or +, others wont let you use those characters but will make sure that you don't use a password that can be looked up as a dictionary word. Some will not allow the same character or number to be used repeatedly. Remember, the more uniqueness you can have in a password makes it that much more difficult for someone to hack your account.

Microsoft Live Account

If you don't already have a Microsoft Live account, you will need to set on up so that you will be able to use Two Factor Authentication. From the work I have done with setting up my MS Live account, you will need to wait about 10 to 15 minutes on average after you make a change before you will see any changes.

Which Two Factor Client to use?

There are several choices to work with for Two Factor Authentication. You can use the MS client. So far, I have only found a client that will install on the Android Operating System. No indication of when a version will be available for the Apple IOS. The Microsoft client pushes out the code that you should see on your Skype client screen. If the codes match, then you can click Approve and you will be in the Skype client momentarily. I have only seen support for the MS software token being used on Microsoft owned properties. Maybe that will change in time but haven't seen anything to indicate a future direction.

If you are only using an Android device, Microsoft has an option that they have released. At this point it is available only on Android. If you select Android, you will be directed to the Google Playstore to download their app.

You have choice of use of 3 different software Two Factor Authentication - Google Authenticator, Authy and SAASPASS. They all do the same basic function in providing you a rotating set of 6 digits to enter when prompted by the application they have been configured for. Authy is a little different in that it synchronizes what you have it configured for so that it backs up for restoring later when needed. I recently became aware of SAASPASS and it has a few other features that set it apart from the competition. It offers a variety of features not found in the other two factor clients I have worked with. One nice feature is that SAASPASS puts an icon to the left of the rotating digits that each particular entry has been configured for.

Configuring your account for two factor use

You start the process by going to https://onedrive.live.com. Login to the account using your username and password. Click on Security & Privacy. Under Account Security, click on More Security Settings. Scroll down the page and look for Two-step verification. It should say Set up two-step verification. Click on that link. At the Set up two-step screen, review the information there and then click on Next to proceed. Since I don't sync my phone with Outlook.com, I clicked on Next to proceed. The next screen you see will show the apps and devices that will need an app password since they won't be able to use a two factor client. Make note if any of these apply so you can followup afterwards. Click on the Finish button.

Using the Microsoft Two Factor Client

Your account is now setup for two step verification. The next thing you will need to do is to select and setup the two factor client to use. Scroll down the screen and look for "Identity Verification Apps". Click on the link "Set up identity verification app". To use the Microsoft two factor client, click on Android (only if you are using an Android phone) and click Next. Once you have downloaded the Microsoft App from the Google Playstore, you will need to setup the account on your smartphone. Just follow the prompts and you should be finished setting up in a few minutes.

Using Google, Authy or SAASPASS two factor clients

At the setup identity screen, select Other as the client type. You will be shown a QR code. Take a screen capture of this and put it in a safe place like an Evernote folder for later use if you have to re-setup your two factor client without having to do a complete re-enroll process. The process, regardless of the two factor client you are using, will be fairly similar. Once you have things setup, you are pretty much ready to go.

Regardless of which two factor client you have elected to use, it will take a few minutes before Skype will be aware of your two factor configuration change. There are a couple of suggestions I would like for you to consider. The first is to enter your cell phone number a as a backup contact method. If for some reason, you can't get the authenticator app to work or the code isn't being accepted for some reason, you can have a text message sent to your phone as a backup authentication step. The other is that while you can have multiple authentication clients configured, don't try that right away. As tempting as it may be, get used to using a single authentication app for this login before you try to get fancy.

Test Two Factor Authentication

- Logout of your Microsoft OneDrive account
- Log back in and test the two factor authentication
- Enter your username and password
- Follow the steps on the screen as indicated for the two factor authentication you have setup

With Microsoft OneDrive, you have the option of using the Two Factor Authentication option you initially setup before using the FIDO U2F key. If you don't have that key available, click Use mobile authenticator option to use that option instead.

Watch Video -
http://protectyourlogin.com/mod

CHAPTER THREE

Domain/Hosting

GoDaddy

GoDaddy is one of the more popular web hosting companies who offer a variety of services. This is another area where you should consider implementing a strong password and two factor authentication to keep you from loosing access to the services you are getting from GoDaddy. Follow the steps outlined here to help prevent that from happening.

Use a Strong Password

For those that aren't familiar with this term, a Strong Password is one that contains (where permitted by the website or service) a combination lower case letters, uppercase letters, numbers, punctuation or other special characters to make up a password. The longer the password is, the long it will take for someone to guess or hack your password. This is an area where I am probably as guilty as everyone else is on this. I have a "favorite" password or two that I liked to use in the past. This kept me from having to have a way to keep track of all the passwords that I used on different websites. With more websites adding additional layers of security such as challenge questions, etc., I have had to start using a password manager app to help keep track of the different passwords, the challenge questions and answers used on a particular website, the recovery procedure if I am locked out of a website, what additional login procedures I have used for that site, etc.

Use a Unique Password

This one will be a bit of a pain but the time taken to do this will pay off in the long run. Look at it this way, if one of your logins on a particular site is compromised or hacked, you have taken a step to minimize the potential for additional website logins to be breached as well. Most password manager apps have the functionality of helping you generate a unique password. There are other apps that do this as well. A search of the app store for your phone or the internet for you laptop/desktop should show you some options to look at.

Different websites have different rules for how you can make your password strong. Some will let you use "special" characters such as * or +, others wont let you use those characters but will make sure that you don't use a password that can be looked up as a dictionary word. Some will not allow the same character or number to be used repeatedly. Remember, the more uniqueness you can have in a password makes it that much more difficult for someone to hack your account.

Enabling Two Factor Authentication

1. Login your GoDaddy Console (https://mya.godaddy.com/default.aspx)
2. Go to the Settings tab.
3. Click Account Security Settings on the left.
4. In the Two-Step Authentication section, click Manage.

6. Click Add Authentication.

7. Enter the smart phone number you want use for the two factor authentication process, and then click Continue.

8. Enter the code we text you, and then click Continue.

There is a little quirk at the end of the process. Once you finish enrolling your phone as the two factor token, you don't get returned to the settings screen where you started from. You will need to click on your login name that should be in the upper right hand corner of the screen and click on the My Account option that will be displayed.

Test Two Factor Authentication

- Logout of your GoDaddy account
- Log back in and test the two factor authentication
- Enter your username and password
- Follow the steps on the screen as indicated for the two factor authentication you have setup

Watch Video -

http://protectyourlogin.com/gd2fa

Hurricane Electric

For those not familiar with Hurricane Electric, HE as they are also known is a DNS Provider and IPv6 Tunnel Broker among other things. If you are using a 3rd party to manage the DNS settings for your domains, enabling two factor authentication is something to consider doing.

Use a Strong Password

For those that aren't familiar with this term, a Strong Password is one that contains (where permitted by the website or service) a combination lower case letters, uppercase letters, numbers, punctuation or other special characters to make up a password. The longer the password is, the long it will take for someone to guess or hack your password. This is an area where I am probably as guilty as everyone else is on this. I have a "favorite" password or two that I liked to use in the past. This kept me from having to have a way to keep track of all the passwords that I used on different websites. With more websites adding additional layers of security such as challenge questions, etc., I have had to start using a password manager app to help keep track of the different passwords, the challenge questions and answers used on a particular website, the recovery procedure if I am locked out of a website, what additional login procedures I have used for that site, etc.

Use a Unique Password

This one will be a bit of a pain but the time taken to do this will pay off in the long run. Look at it this way, if one of your logins on a particular site is compromised or hacked, you have taken a step to minimize the potential for additional website logins to be breached as well. Most password manager apps have the functionality of helping you generate a unique password. There are other apps that do this as well. A search of the app store for your phone or the internet for you laptop/desktop should show you some options to look at.

Different websites have different rules for how you can make your password strong. Some will let you use "special" characters such as * or +, others wont let you use those characters but will make sure that you don't use a password that can be looked up as a dictionary word. Some will not allow the same character or number to be used repeatedly. Remember, the more uniqueness you can have in a password makes it that much more difficult for someone to hack your account.

Enabling Two Factor Authentication

- Login to your HE account.
- Go to the Account Information page.
- Look for Two Factor section on page.
- Click on Enable button.
- Do a screen capture of the QR code and put it in a safe place.
- Enroll the account in your Two Factor Authentication client of choice

- Enter the current code displayed on the Two Factor Authentication client
- Click enable

Test Two Factor Authentication

- Logout of your Hurricane Electric account
- Log back in and test the two factor authentication
- Enter your username and password
- Follow the steps on the screen as indicated for the two factor authentication you have setup

With Hurricane Electric, you have the option of using the Two Factor Authentication option you initially setup before using the FIDO U2F key. If you don't have that key available, click Use mobile authenticator option to use that option instead.

Watch Video -

http://protectyourlogin.com/he2fa

Simple Niche Domains

Simple Niche Domains is a provider of DNS and Web Hosting services. This is another area where you should consider implementing a strong password and two factor authentication to keep you from loosing access to the services. Follow the steps outlined here to help prevent that from happening..

Use a Strong Password

For those that aren't familiar with this term, a Strong Password is one that contains (where permitted by the website or service) a combination lower case letters, uppercase letters, numbers, punctuation or other special characters to make up a password. The longer the password is, the long it will take for someone to guess or hack your password. This is an area where I am probably as guilty as everyone else is on this. I have a "favorite" password or two that I liked to use in the past. This kept me from having to have a way to keep track of all the passwords that I used on different websites. With more websites adding additional layers of security such as challenge questions, etc., I have had to start using a password manager app to help keep track of the different passwords, the challenge questions and answers used on a particular website, the recovery procedure if I am locked out of a website, what additional login procedures I have used for that site, etc.

Use a Unique Password

This one will be a bit of a pain but the time taken to do this will pay off in the long run. Look at it this way, if one of your logins on a particular site is compromised or hacked, you have taken a step to minimize the potential for additional website logins to be breached as well. Most password manager apps have the functionality of helping you generate a unique password. There are other apps that do this as well. A search of the app store for your phone or the internet for you laptop/desktop should show you some options to look at.

Different websites have different rules for how you can make your password strong. Some will let you use "special" characters such as * or +, others wont let you use those characters but will make sure that you don't use a password that can be looked up as a dictionary word. Some will not allow the same character or number to be used repeatedly. Remember, the more uniqueness you can have in a password makes it that much more difficult for someone to hack your account.

Enabling Two Factor Authentication

1. Login your Simple Niche Domains Console (http://www.simplenichedomains.com)

2. Click on the settings button

2. Click on Account Security Settings tab.

3. Click on Manage in the 2-Step Authentication window

4. When the Two Step Authentication window

appears, click on Add Authentication

5. Enter the phone number for your Smart Phone

6. Click on Continue

7. Enter the 6 digit code you will receive in the box on the screen

8. Click on the Continue button

There is a little quirk at the end of the process. Once you finish enrolling your phone as the two factor token, you don't get returned to the settings screen where you started from. You will need to click on your login name that should be in the upper right hand corner of the screen and click on the My Account option that will be displayed.

Test Two Factor Authentication

- Logout of your Simple Niche Domains account
- Log back in and test the two factor authentication
- Enter your username and password
- Follow the steps on the screen as indicated for the two factor authentication you have setup

With Simple Niche Domains, you have the option of using the Two Factor Authentication option you initially setup before using the FIDO U2F key. If you don't have that key available, click Use mobile authenticator option to use that option instead.

Watch Video -

http://protectyourlogin.com/snd

CHAPTER FOUR

Communications

Gmail

Use a Strong Password

For those that aren't familiar with this term, a Strong Password is one that contains (where permitted by the website or service) a combination lower case letters, uppercase letters, numbers, punctuation or other special characters to make up a password. The longer the password is, the long it will take for someone to guess or hack your password. This is an area where I am probably as guilty as everyone else is on this. I have a "favorite" password or two that I liked to use in the past. This kept me from having to have a way to keep track of all the passwords that I used on different websites. With more websites adding additional layers of security such as challenge questions, etc., I have had to start using a password manager app to help keep track of the different passwords, the challenge questions and answers used on a particular website, the recovery procedure if I am locked out of a website, what additional login procedures I have used for that site, etc.

Use a Unique Password

This one will be a bit of a pain but the time taken to do this will pay off in the long run. Look at it this way, if one of your logins on a particular site is compromised or hacked, you have taken a step to minimize the potential for additional website logins to be breached as well. Most password manager apps have the functionality of

helping you generate a unique password. There are other apps that do this as well. A search of the app store for your phone or the internet for you laptop/desktop should show you some options to look at.

Different websites have different rules for how you can make your password strong. Some will let you use "special" characters such as * or +, others wont let you use those characters but will make sure that you don't use a password that can be looked up as a dictionary word. Some will not allow the same character or number to be used repeatedly. Remember, the more uniqueness you can have in a password makes it that much more difficult for someone to hack your account.

Enabling Two Factor Authentication

- Login to your Google account
- Click on your account name in the upper right hand corner of the screen
- Click on My Account
- Click on Sign-in & Security
- Click on 2 Step Verification
- Click on Start Setup
- Enter number of your smart phone
- Select SMS
- Click on Send Code
- Enter code received on your smart phone and click on Verify
- Click on Next to trust the computer you are working on
- Click on Confirm to turn on 2 Step Verification

- You should now be at 2-Step Verification screen

To use the Yubico U2F Key

- Click on Security Keys tab
- Click on Add Security Key
- Click on Register
- Insert key and tap on button when it starts flashing
- Click on Done when the key is registered

To use a Soft Token (This can be used for the Yubikey NEO as well)

- Click on Switch to app
- Click on the type of smart phone you are using
- Click on Continue
- Take a screen capture of the current QR code
- Enroll the QR code into the soft token app that you are using
- Enter the 6 digit code displayed by the phone
- Click on Verify
- You should now get a verified screen
- Token is enrolled

Another way of getting back to the token enrollment screen is to go to https://www.google.com/landing/2step/.

A part of the setup process is to print the backup codes as an option to get into the account (just in case). These should be put in a safe place in case you misplace your token/smartphone and need to get into the account. Create a folder in

Evernote and file the QR code and the backup codes for your two factor accounts. If you have any apps that need to login to your Google Accounts and don't understand two factor authentication, you will want to set up an application password (a very long and very difficult password) so that the application can get into your Google account.

Test Two Factor Authentication

- Logout of your Gmail account
- Log back in and test the two factor authentication
- Enter your username and password
- Follow the steps on the screen as indicated for the two factor authentication you have setup

With Gmail, you have the option of using the Two Factor Authentication option you initially setup before using the FIDO U2F key. If you don't have that key available, click Use mobile authenticator option to use that option instead.

Watch Video -

http://protectyourlogin.com/go2fa

Skype

Skype is a popular account for doing video chats or interviews with the media or podcasts. It should be protected just like any other online account that you have. By following the steps I have outlined, you will do as much as you can to keep others from trying to get into your account.

Use a Strong Password

For those that aren't familiar with this term, a Strong Password is one that contains (where permitted by the website or service) a combination lower case letters, uppercase letters, numbers, punctuation or other special characters to make up a password. The longer the password is, the long it will take for someone to guess or hack your password. This is an area where I am probably as guilty as everyone else is on this. I have a "favorite" password or two that I liked to use in the past. This kept me from having to have a way to keep track of all the passwords that I used on different websites. With more websites adding additional layers of security such as challenge questions, etc., I have had to start using a password manager app to help keep track of the different passwords, the challenge questions and answers used on a particular website, the recovery procedure if I am locked out of a website, what additional login procedures I have used for that site, etc.

Use a Unique Password

This one will be a bit of a pain but the time taken to do this will pay off in the long run. Look at it this way, if one of your logins on a particular site is compromised or hacked, you have taken a step to minimize the potential for additional website logins to be breached as well. Most password manager apps have the functionality of helping you generate a unique password. There are other apps that do this as well. A search of the app store for your phone or the internet for you laptop/desktop should show you some options to look at.

Different websites have different rules for how you can make your password strong. Some will let you use "special" characters such as * or +, others wont let you use those characters but will make sure that you don't use a password that can be looked up as a dictionary word. Some will not allow the same character or number to be used repeatedly. Remember, the more uniqueness you can have in a password makes it that much more difficult for someone to hack your account.

Microsoft Live or Skype Account

If you don't already have a Microsoft Live account, you will need to set on up so that you will be able to use Two Factor Authentication. From the work I have done with setting up my MS Live account, you will need to wait about 10 to 15 minutes on average after you make a change before you will see any changes with the Skype client. I have noticed a "feature" in the smartphone Skype client that you might want to be aware of. Once you have logged in one time to the client, it will remember it and not prompt you for two factor authentication again which somewhat defeats the purpose of two factor authentication. If you select the login with another account option, you will be prompted for two factor authentication. I have been able to reproduce this situation. Haven't been able to work around it.

Which Two Factor Client to use?

There are several choices to work with for Two Factor Authentication. You can use the MS client. So far, I have only found a client that will install on the Android Operating System. No indication of when a version will be available for the Apple IOS. The Microsoft client pushes out the code that you should see on your Skype client screen. If the codes match, then you can click Approve and you will be in the Skype client momentarily. I have only seen support for the MS software token being used on Microsoft owned properties. Maybe that will change in time but haven't seen anything to indicate a future direction.

If you are only using an Android device, Microsoft has an option that they have released. At this point it is available only on Android. If you select Android, you will be directed to the Google Playstore to download their app.

You have choice of use of 3 different software Two Factor Authentication - Google Authenticator, Authy and SAASPASS. They all do the same basic function in providing you a rotating set of 6 digits to enter when prompted by the application they have been configured for. Authy is a little different in that it synchronizes what you have it configured for so that it backs up for restoring later when needed. I recently became aware of SAASPASS and it has a few other features that set it apart from the competition. It offers a variety of features not found in the other two factor clients I have worked with. One nice feature is that SAASPASS puts an icon to the left of the rotating digits that each particular entry has been configured for.

Configuring your account for two factor use

You start the process by going to https://login.live.com. Login to the account using your username and password. Click on Security & Privacy. Under Account Security, click on More Security Settings. Scroll down the page and look for Two-step verification. It should say Set up two-step verification. Click on that link. At the Set up two-step screen, review the information there and then click on Next to proceed. Since I don't sync my phone with Outlook.com, I clicked on Next to proceed. The next screen you see will show the apps and devices that will need an app password since they won't be able to use a two factor client. Make note if any of these apply so you can followup afterwards. Click on the Finish button.

Using the Microsoft Two Factor Client

Your account is now setup for two step

verification. The next thing you will need to do is to select and setup the two factor client to use. Scroll down the screen and look for "Identity Verification Apps". Click on the link "Set up identity verification app". To use the Microsoft two factor client, click on Android (only if you are using an Android phone) and click Next. Once you have downloaded the Microsoft App from the Google Playstore, you will need to setup the account on your smartphone. Just follow the prompts and you should be finished setting up in a few minutes.

Using Google, Authy or SAASPASS two factor clients

At the setup identity screen, select Other as the client type. You will be shown a QR code. Take a screen capture of this and put it in a safe place like an Evernote folder for later use if you have to re-setup your two factor client without having to do a complete re-enroll process. The process, regardless of the two factor client you are using, will be fairly similar. Once you have things setup, you are pretty much ready to go.

Regardless of which two factor client you have elected to use, it will take a few minutes before Skype will be aware of your two factor configuration change. There are a couple of suggestions I would like for you to consider. The first is to enter your cell phone number a as a backup contact method. If for some reason, you can't get the authenticator app to work or the code isn't being accepted for some reason, you can have a text message sent to your phone as a backup authentication step. The other is that while you can have multiple authentication clients configured, don't try that right away. As tempting as it may be, get used to using a single authentication app for this login before you try to get fancy.

Test Two Factor Authentication

- Logout of your Skype account
- Log back in and test the two factor authentication
- Enter your username and password
- Follow the steps on the screen as indicated for the two factor authentication you have setup

With Skype, you have the option of using the Two Factor Authentication option you initially setup before using the FIDO U2F key. If you don't have that key available, click Use mobile authenticator option to use that option instead.

Watch Video -

http://protectyourlogin.com/sk2fa

Yahoo

Use a Strong Password

For those that aren't familiar with this term, a Strong Password is one that contains (where permitted by the website or service) a combination lower case letters, uppercase letters, numbers, punctuation or other special characters to make up a password. The longer the password is, the long it will take for someone to guess or hack your password. This is an area where I am probably as guilty as everyone else is on this. I have a "favorite" password or two that I liked to use in the past. This kept me from having to have a way to keep track of all the passwords that I used on different websites. With more websites adding additional layers of security such as challenge questions, etc., I have had to start using a password manager app to help keep track of the different passwords, the challenge questions and answers used on a particular website, the recovery procedure if I am locked out of a website, what additional login procedures I have used for that site, etc.

Use a Unique Password

This one will be a bit of a pain but the time taken to do this will pay off in the long run. Look at it this way, if one of your logins on a particular site is compromised or hacked, you have taken a step to minimize the potential for additional website logins to be breached as well. Most password manager apps have the functionality of

helping you generate a unique password. There are other apps that do this as well. A search of the app store for your phone or the internet for you laptop/desktop should show you some options to look at.

Different websites have different rules for how you can make your password strong. Some will let you use "special" characters such as * or +, others wont let you use those characters but will make sure that you don't use a password that can be looked up as a dictionary word. Some will not allow the same character or number to be used repeatedly. Remember, the more uniqueness you can have in a password makes it that much more difficult for someone to hack your account.

Enabling Two Factor Authentication

- Login to Yahoo account
- Click on Name
- Click on Account Info
- Click on Account Security
- Click on Two Factor Authentication
- Enter phone number to call
- Click Send SMS
- Enter code sent and click on Verify button

Application Passwords

There are times such as when you link your Yahoo account with something like IFTTT and using 2FA (Two Factor Authentication) won't be an option. This is where an application password will be useful. You will want to generate a different

Application Password for each situation like this that you have. In that way, if one app does get compromised, you have the ability to block access from that app without having to redo your authentication setup. Click Skip for now and you can set this up later if needed.

Test Two Factor Authentication

- Logout of your Yahoo account
- Log back in and test the two factor authentication
- Enter your username and password
- Follow the steps on the screen as indicated for the two factor authentication you have setup

With Yahoo, you have the option of using the Two Factor Authentication option you initially setup before using the FIDO U2F key. If you don't have that key available, click Use mobile authenticator option to use that option instead.

Watch Video -

https://youtu.be/5ydftj7nXBY

CHAPTER FIVE

Other

Social Security Administration

This is probably not an account that a lot of people will have. Because of the nature of the information that is available, you should get one set up as soon as possible. The instructions given here will assume that you are setting the account up from scratch.

Use a Strong Password

For those that aren't familiar with this term, a Strong Password is one that contains (where permitted by the website or service) a combination lower case letters, uppercase letters, numbers, punctuation or other special characters to make up a password. The longer the password is, the long it will take for someone to guess or hack your password. This is an area where I am probably as guilty as everyone else is on this. I have a "favorite" password or two that I liked to use in the past. This kept me from having to have a way to keep track of all the passwords that I used on different websites. With more websites adding additional layers of security such as challenge questions, etc., I have had to start using a password manager app to help keep track of the different passwords, the challenge questions and answers used on a particular website, the recovery procedure if I am locked out of a website, what additional login procedures I have used for that site, etc.

Use a Unique Password

This one will be a bit of a pain but the time taken to do this will pay off in the long run. Look at it this way, if one of your logins on a particular site is compromised or hacked, you have taken a step to minimize the potential for additional website logins to be breached as well. Most password manager apps have the functionality of helping you generate a unique password. There are other apps that do this as well. A search of the app store for your phone or the internet for you laptop/desktop should show you some options to look at.

Different websites have different rules for how you can make your password strong. Some will let you use "special" characters such as * or +, others wont let you use those characters but will make sure that you don't use a password that can be looked up as a dictionary word. Some will not allow the same character or number to be used repeatedly. Remember, the more uniqueness you can have in a password makes it that much more difficult for someone to hack your account.

Creating your Social Security Administration login

1. When you are creating the account, make sure that you select the Yes, let's start now option under Add Extra Security.

2. You will need to select one of four options as to how your information will be verified

3. Enter the information requested based on the option that you selected for verifying your identity

4. Click the Next button

5. You may be asked a series of questions to verify who you are. Select the answers for the questions given. Click on the Next button

6. Enter a username for this login

7. Enter the password for this login

8. Enter the email address to be used for this account

9. Re-enter the password for this account

10. Select the questions and supply the answer for 3 security questions

11. Record these questions and answers in your password manager app

12. Click the Next button

13. You should now get a screen that says your account has been created. You will receive an upgrade code in the mail with instructions on how to upgrade your account to use the extra security you selected when creating the account.

14. Click on the Next button

15. Review the terms of service

16. Check the box beside "I agree to the Terms of Service"

17. Click on Next

When you receive the upgrade code in the mail

1. Go to www.socialsecurity.gov/myaccount

2. Click on Sign In

3. Enter Your Username and Password as indicated on the screen

4. Click on the Sign In button

5. Click on Yes, enable my extra security

6. Click on Next
7. Enter Cell phone number where indicated
8. Click on the Get Text Message button
9. Enter the code you received via SMS and click Next
10. Get the one-time Account Upgrade Code from your letter and enter it on the screen where indicated
11. Click on Next
12. When you see the "Congratulations" screen, you code has been accepted
13. Click on Next
14. Review the Terms of Service
15. Check the box beside "I agree to the Terms of Service"
16. Click on the Next button

Two Factor Authentication is now enabled on your social security account !

If you do use Google voice as your cell phone number, you may need to check your spam folder when looking for the SMS message. My Google voice SMS messages don't normally end up in this folder but this particular message did.

Watch Video -
http://protectyourlogin.com/ss2fa

CHAPTER SIX

Remote Access

LogMeIn

Use a Strong Password

For those that aren't familiar with this term, a Strong Password is one that contains (where permitted by the website or service) a combination lower case letters, uppercase letters, numbers, punctuation or other special characters to make up a password. The longer the password is, the long it will take for someone to guess or hack your password. This is an area where I am probably as guilty as everyone else is on this. I have a "favorite" password or two that I liked to use in the past. This kept me from having to have a way to keep track of all the passwords that I used on different websites. With more websites adding additional layers of security such as challenge questions, etc., I have had to start using a password manager app to help keep track of the different passwords, the challenge questions and answers used on a particular website, the recovery procedure if I am locked out of a website, what additional login procedures I have used for that site, etc.

Use a Unique Password

This one will be a bit of a pain but the time taken to do this will pay off in the long run. Look at it this way, if one of your logins on a particular site is compromised or hacked, you have taken a step to minimize the potential for additional website logins to be breached as well. Most password manager apps have the functionality of

helping you generate a unique password. There are other apps that do this as well. A search of the app store for your phone or the internet for you laptop/desktop should show you some options to look at.

Different websites have different rules for how you can make your password strong. Some will let you use "special" characters such as * or +, others wont let you use those characters but will make sure that you don't use a password that can be looked up as a dictionary word. Some will not allow the same character or number to be used repeatedly. Remember, the more uniqueness you can have in a password makes it that much more difficult for someone to hack your account.

Enabling Two Factor Authentication

- Login to your account
- Click on Settings
- Click on Security
- Check the box on emailed security code
- Enter the email address the login code should be sent to
- Click on Save
- Enter the security code emailed to you in the box on the screen
- Click on Ok

Two Factor Authentication is now setup

Watch Video -
http://protectyourlogin.com/lmi2fa

TeamViewer

Use a Strong Password

For those that aren't familiar with this term, a Strong Password is one that contains (where permitted by the website or service) a combination lower case letters, uppercase letters, numbers, punctuation or other special characters to make up a password. The longer the password is, the long it will take for someone to guess or hack your password. This is an area where I am probably as guilty as everyone else is on this. I have a "favorite" password or two that I liked to use in the past. This kept me from having to have a way to keep track of all the passwords that I used on different websites. With more websites adding additional layers of security such as challenge questions, etc., I have had to start using a password manager app to help keep track of the different passwords, the challenge questions and answers used on a particular website, the recovery procedure if I am locked out of a website, what additional login procedures I have used for that site, etc.

Use a Unique Password

This one will be a bit of a pain but the time taken to do this will pay off in the long run. Look at it this way, if one of your logins on a particular site is compromised or hacked, you have taken a step to minimize the potential for additional website logins to be breached as well. Most password manager apps have the functionality of

helping you generate a unique password. There are other apps that do this as well. A search of the app store for your phone or the internet for you laptop/desktop should show you some options to look at.

Different websites have different rules for how you can make your password strong. Some will let you use "special" characters such as * or +, others wont let you use those characters but will make sure that you don't use a password that can be looked up as a dictionary word. Some will not allow the same character or number to be used repeatedly. Remember, the more uniqueness you can have in a password makes it that much more difficult for someone to hack your account.

Enabling Two Factor Authentication

- Login to your account
- Click on your name
- Click on Edit Profile
- Click on Activate beside Two Factor Authentication
- Click on Start Activation
- Do a screen capture when the QR code appears on the screen
- Put this in a folder in Evernote or whatever app you are using to keep backups copies of the QR code
- Enroll the QR code in the Two Factor Client you are using
- Click on Next
- Enter the code current displayed by the 2FA

client

- Click on Activate
- Copy the recovery code displayed or do a print screen and place it in Evernote
- Click on Close
- Click on Save

Watch Video -
http://protectyourlogin.com/tv2fa

CHAPTER SEVEN

Retail

eBay

Use a Strong Password

For those that aren't familiar with this term, a Strong Password is one that contains (where permitted by the website or service) a combination lower case letters, uppercase letters, numbers, punctuation or other special characters to make up a password. The longer the password is, the long it will take for someone to guess or hack your password. This is an area where I am probably as guilty as everyone else is on this. I have a "favorite" password or two that I liked to use in the past. This kept me from having to have a way to keep track of all the passwords that I used on different websites. With more websites adding additional layers of security such as challenge questions, etc., I have had to start using a password manager app to help keep track of the different passwords, the challenge questions and answers used on a particular website, the recovery procedure if I am locked out of a website, what additional login procedures I have used for that site, etc.

Use a Unique Password

This one will be a bit of a pain but the time taken to do this will pay off in the long run. Look at it this way, if one of your logins on a particular site is compromised or hacked, you have taken a step to minimize the potential for additional website logins to be breached as well. Most password manager apps have the functionality of

helping you generate a unique password. There are other apps that do this as well. A search of the app store for your phone or the internet for you laptop/desktop should show you some options to look at.

Different websites have different rules for how you can make your password strong. Some will let you use "special" characters such as * or +, others wont let you use those characters but will make sure that you don't use a password that can be looked up as a dictionary word. Some will not allow the same character or number to be used repeatedly. Remember, the more uniqueness you can have in a password makes it that much more difficult for someone to hack your account.

Security Questions

If you haven't logged into your eBay account lately, you will be prompted to answer a series of security questions. If you have already entered these and have written down both the question and the answer you provided, that's great. If you don't remember that you entered, go through the questions again and enter into the note field of your Password Manager app both the questions and the answers.

Enabling Two Factor Authentication

You have the option of either using one of the following security keys with your eBay account - Symantec VIP Soft Token (available in either Apple or Google Play stores)or the Yubico VIP key. The instructions here are for the Yubico VIP key.

- Login to your eBay Account
- Click on your name
- Click on Account Settings
- Click on Account
- Click on Personal Information
- Scroll down to Security Information
- Click on the Activation link
- Enter the serial number on the back of your Yubikey VIP
- Insert the key in a USB port on your computer
- Click in the field for the first 6 digit security code
- Tap on the button on the Yubikey VIP key
- Click in the field for the 2nd 6 digit security code
- Tap on the button on the Yubikey VIP key
- Click on the Activate button

You should be momentarily returned back to the Personal Information screen. You should see a message just under the Personal Information bar indicating that you have successfully activated a Security Key for your eBay account.

Test Two Factor Authentication

- Logout of your eBay account
- Log back in and test the two factor authentication
- Enter your username and password
- Follow the steps on the screen as indicated for

the two factor authentication you have setup

Watch Video -
http://protectyourlogin.com/eBay2FA

PayPal

Use a Strong Password

For those that aren't familiar with this term, a Strong Password is one that contains (where permitted by the website or service) a combination lower case letters, uppercase letters, numbers, punctuation or other special characters to make up a password. The longer the password is, the long it will take for someone to guess or hack your password. This is an area where I am probably as guilty as everyone else is on this. I have a "favorite" password or two that I liked to use in the past. This kept me from having to have a way to keep track of all the passwords that I used on different websites. With more websites adding additional layers of security such as challenge questions, etc., I have had to start using a password manager app to help keep track of the different passwords, the challenge questions and answers used on a particular website, the recovery procedure if I am locked out of a website, what additional login procedures I have used for that site, etc.

Use a Unique Password

This one will be a bit of a pain but the time taken to do this will pay off in the long run. Look at it this way, if one of your logins on a particular site is compromised or hacked, you have taken a step to minimize the potential for additional website logins to be breached as well. Most password manager apps have the functionality of

helping you generate a unique password. There are other apps that do this as well. A search of the app store for your phone or the internet for you laptop/desktop should show you some options to look at.

Different websites have different rules for how you can make your password strong. Some will let you use "special" characters such as * or +, others wont let you use those characters but will make sure that you don't use a password that can be looked up as a dictionary word. Some will not allow the same character or number to be used repeatedly. Remember, the more uniqueness you can have in a password makes it that much more difficult for someone to hack your account.

Security Questions

If you haven't logged into your eBay account lately, you will be prompted to answer a series of security questions. If you have already entered these and have written down both the question and the answer you provided, that's great. If you don't remember that you entered, go through the questions again and enter into the note field of your Password Manager app both the questions and the answers.

Enabling Two Factor Authentication

You have the option of either using one of the following security keys with your eBay account - Symantec VIP Soft Token (available in either Apple or Google Play stores)or the Yubico VIP key, or use your smart phone as a token. You can enter more than one key, you should be prompted for which key to use when you login. Try starting with one key first before bringing a second key into the mix.

1) Registering your smart phone as a Two Factor Token

- Login to your PayPal Account
- Click on the settings icon (looks like a gear) in the upper right hand corner of the page
- Click on Security (should be just under your name
- Click on Edit to the right of the Security Key label
- Click on Get Security Key at the bottom of the screen
- Enter your smart phone number in the designated areas of the screen
- Click on the Agree and register
- Enter the 6 digit code you that arrived on the phone
- Your phone is now registered as a Two Factor token

2) Registering your Yubico VIP as a Two Factor Token

- Login to your eBay Account
- Click on your name
- Click on Account Settings
- Click on Account
- Click on Personal Information
- Scroll down to Security Information

- Click on the PayPal Activations link
- Enter the serial number on the back of your Yubikey VIP
- Insert the key in a USB port on your computer
- Click in the field for the first 6 digit security code
- Tap on the button on the Yubikey VIP key
- Click in the field for the 2nd 6 digit security code
- Tap on the button on the Yubikey VIP key
- Click on the Activate button

Test Two Factor Authentication

- Logout of your PayPal account
- Log back in and test the two factor authentication
- Enter your username and password
- Follow the steps on the screen as indicated for the two factor authentication you have setup

Watch Video -

http://protectyourlogin.com/pay2fa

Amazon

Amazon may not seem like a choice to immediately get hardened. When you look at the information that is kept on this particular site - credit cards used, the addresses that items have been shipped to and what you have purchased could end up being a gold mine for someone trying to impersonate you to get additional credit in your name but for their benefit.

Use a Strong Password

For those that aren't familiar with this term, a Strong Password is one that contains (where permitted by the website or service) a combination lower case letters, uppercase letters, numbers, punctuation or other special characters to make up a password. The longer the password is, the long it will take for someone to guess or hack your password. This is an area where I am probably as guilty as everyone else is on this. I have a "favorite" password or two that I liked to use in the past. This kept me from having to have a way to keep track of all the passwords that I used on different websites. With more websites adding additional layers of security such as challenge questions, etc., I have had to start using a password manager app to help keep track of the different passwords, the challenge questions and answers used on a particular website, the recovery procedure if I am locked out of a website, what additional login procedures I have used for that site, etc.

Use a Unique Password

This one will be a bit of a pain but the time taken to do this will pay off in the long run. Look at it this way, if one of your logins on a particular site is compromised or hacked, you have taken a step to minimize the potential for additional website logins to be breached as well. Most password manager apps have the functionality of helping you generate a unique password. There are other apps that do this as well. A search of the app store for your phone or the internet for you laptop/desktop should show you some options to look at.

Different websites have different rules for how you can make your password strong. Some will let you use "special" characters such as * or +, others wont let you use those characters but will make sure that you don't use a password that can be looked up as a dictionary word. Some will not allow the same character or number to be used repeatedly. Remember, the more uniqueness you can have in a password makes it that much more difficult for someone to hack your account.

Enabling Two Factor Authentication

- Login to Amazon account
- Click on "Your Account"
- Scroll down the List and click on the 2nd "Your Account" item listed
- From the Your Account Page, click on Change

Account Settings option under Settings, Account Settings

- Click on Advanced Security Settings
- At the Two Step Verification page, click on the Get Started button
- Enter your cell phone number where directed on the screen
- Click On the Send code button
- When the Add backup method screen appears, Select Authenticator app
- Do a screen capture of the displayed barcode
- Scan the barcode shown on the screen using your Authenticator app
- Enter the value from the authenticator app you are using
- Click on Verify
- When How to use Two-Step Verification screen appears, click on the Got it button at the bottom of the screen

This configuration will go to the cell phone first, then the authenticator app, you can change the order at any time.

Watch Video -

http://protectyourlogin.com/Az2FA

CHAPTER EIGHT

Social Media

Blogger

Since this is a Google property. You can use your Google Credentials to login. If you don't have a Google account, you can setup an account by clicking on this link to Setup up Gmail.

Watch Video -

http://protectyourlogin.com/go2fa

Facebook

A good friend of mine recently had their Facebook account hacked and it took several days to get things back under control. Even though I had taken some steps to try to prevent this from happening to me, I decided it was a good time to review what I had done and see if there were any improvements that I could make. I did find a few things to do and have taken those steps to increase the level of protection around my account. Here is a review of what you can look to do to prevent your Facebook account from being taken over by someone else. Don't feel that you have to implement these all at the same time. Gradually putting these steps into affect will help you get used to the change in the login process. The only thing that is worse than only doing one or two of the steps that I will outline will be to do nothing at all. All of the settings I refer to can be found under your Profile on Facebook under Settings, Security Settings.

Use a Strong Password

For those that aren't familiar with this term, a Strong Password is one that contains (where permitted by the website or service) a combination lower case letters, uppercase letters, numbers, punctuation or other special characters to make up a password. The longer the password is, the long it will take for someone to guess or hack your password. This is an area where I am probably as guilty as everyone else is on this. I have a "favorite" password or two that I liked to use in the past. This kept me from having to have a way to keep track of all the passwords that I used on different websites. With more websites adding additional layers of security such as challenge questions, etc., I have had to start using a password manager app to help keep track of the different passwords, the challenge questions and answers used on a particular website, the recovery procedure if I am locked out of a website, what additional login procedures I have used for that site, etc.

Use a Unique Password

This one will be a bit of a pain but the time taken to do this will pay off in the long run. Look at it this way, if one of your logins on a particular site is compromised or hacked, you have taken a step to minimize the potential for additional website logins to be breached as well. Most password manager apps have the functionality of helping you generate a unique password. There are other apps that do this as well. A search of the app store for your phone or the internet for you laptop/desktop should show you some options to look at.

Different websites have different rules for how you can make your password strong. Some will let you use "special" characters such as * or +, others wont let you use those characters but will make sure that you don't use a password that can be looked up as a dictionary word. Some will not allow the same character or number to be used repeatedly. Remember, the more uniqueness you can have in a password makes it that much more difficult for someone to hack your account.

Login Approval from Cellphone

This is an option under the security settings in your Facebook account that will require the entry of a 6 digit code sent by Facebook via SMS to the cell associated with your user account. If you can enabled the two factor authentication described later, this setting wont be enforced as you are using something that is considered more secure at protecting your account. If you don't want to implement two factor login or aren't comfortable with doing so at this point in time, this is an option worth looking at as an additional precautionary measure. I have seen one exception to this after I enabled it and cleared all of the open or active sessions, that I get prompted for a SMS code instead of a pin from the Google authenticator app when logging in using the Facebook app on my smartphone.

Login Alerts

You can have FaceBook send you an email or text when a login is attempted from a browser that isn't recognized. Even with having two factor authentication setup on my account, this is an additional level of notification just to remind me that I will get notified when logging in as a reminder to check if I get this kind of email and I haven't logged in to check for a problem on the account.

Code Generator

This could be worded a little differently as it initially doesn't make a lot of sense. Google created a two factor authentication service several years ago. If you haven't heard of two factor authentication before, look at it this way, a password is something you know, the code from a login token app is something you have - hence the name two factor authentication. Using the Google Authenticator App and your smart phone, you establish a link to your Facebook account that during the login process, you can enter a 6 digit pin that is constantly changing from the Google Authenticator App that increases the level of difficulty of someone logging in as you. You can find this in the app store for you Android or iPhone Smartphone by searching for "Google Authenticator".

During the registration/setup process for the Google Authenticator app or whatever compatible app you are using (such as Authy), there is what is known as a QR code that will be displayed on the screen. You will use the camera on your smartphone to "read" the QR code. This gets the information into the app you are using for two factor authentication without having to enter a long string of numbers and letters. Since other social media and websites are also starting to use this two factor system, I usually take a screenshot of each QR code that I have used to setup the two factor app on my phone so when I change phones, and I do, that the process of setting up the two factor app on the new phone will be a little less painful.

If you don't have an authenticator app loaded on the smart phone you are using to access Facebook, or are accessing Facebook from a desktop/laptop, you still can use Two Factor Authentication. Facebook put this functionality in their smart phone apps. If you don't have a popup message on your phone, you should be taken to a Code Generator screen when you bring up the Facebook app on your smartphone. It is always nice to have a backup plan.

I have seen an interesting change with the last Facebook client for smartphones. When you log in with an account setup for Two Factor Authentication, a screen labeled Code Generator will appear on the screen of your smartphone/ tablet or you should get a notification that when tapped should bring up the Code Generator screen from the Facebook app. You have the option of using this code or the one generated by the Google Authenticator/Authy/SAASPASS soft token that you elected to use.

Remember Browser screen prompt

This is a question that you will get during the login process where you can "remember" the browser that you are logging in from so you don't have to use all of the additional login steps that you have just put in place. I would suggest answer no to this one every time you log in. This may be a pain but look at it this way - if your mobile device or laptop is stolen, you have just added an addition level of protection by requiring that the two factor authentication process be used on every login attempt. If you do happen to Remember the browser you are using, that will supersede the use of two factor authentication for this browser. You can get that to work again by following what is suggested in the next step.

Delete all active browser and app sessions that Facebook knows about

I initially didn't do this and wondered why that the additional levels of security I had enabled weren't taking affect. The best plans in the world aren't any good unless you make sure that they are actually working. This is a setting that you will need to periodically review to make sure an errant session hasn't made its way into the list and therefore make it easier for someone else to get access to your account.

Logout of Facebook

This is another area that I am guilty of as well. When in a hurry, I just close the app or browser session instead of logging out. Taking just a second or two to use the logout process might be the one thing that could make it more difficult for someone to login to your account the next time. I hope that these steps help you keep your Facebook account safe from unwelcome visitors.

Test Two Factor Authentication

- Logout of your Facebook account
- Log back in and test the two factor authentication
- Enter your username and password
- Follow the steps on the screen as indicated for the two factor authentication you have setup

With Facebook, you have the option of using the Two Factor Authentication option you initially setup before using the FIDO U2F key. If you don't have that key available, click Use mobile authenticator option to use that option instead.

Watch Video -

http://protectyourlogin.com/FB2FA

Instagram

Use a Strong Password

For those that aren't familiar with this term, a Strong Password is one that contains (where permitted by the website or service) a combination lower case letters, uppercase letters, numbers, punctuation or other special characters to make up a password. The longer the password is, the long it will take for someone to guess or hack your password. This is an area where I am probably as guilty as everyone else is on this. I have a "favorite" password or two that I liked to use in the past. This kept me from having to have a way to keep track of all the passwords that I used on different websites. With more websites adding additional layers of security such as challenge questions, etc., I have had to start using a password manager app to help keep track of the different passwords, the challenge questions and answers used on a particular website, the recovery procedure if I am locked out of a website, what additional login procedures I have used for that site, etc.

Use a Unique Password

This one will be a bit of a pain but the time taken to do this will pay off in the long run. Look at it this way, if one of your logins on a particular site is compromised or hacked, you have taken a step to minimize the potential for additional website logins to be breached as well. Most password manager apps have the functionality of

helping you generate a unique password. There are other apps that do this as well. A search of the app store for your phone or the internet for you laptop/desktop should show you some options to look at.

Different websites have different rules for how you can make your password strong. Some will let you use "special" characters such as * or +, others wont let you use those characters but will make sure that you don't use a password that can be looked up as a dictionary word. Some will not allow the same character or number to be used repeatedly. Remember, the more uniqueness you can have in a password makes it that much more difficult for someone to hack your account.

Enabling Two Factor Authentication

Instagram is one of those services that doesn't support any of the two factor systems that services such as Facebook or Dropbox do for an additional layer of protection. There is an option. You can use the steps outlined in the chapter "For Sites that don't support Two Factor"

LinkedIn

Along with Facebook, LinkedIn is another popular social media site. It is also a prime candidate for having an account compromised. As with any social media or website, there are several things that can be done to make it a less than appealing target and encourage the would be attacker to move to another less prepared account.

Use a Strong Password

For those that aren't familiar with this term, a Strong Password is one that contains (where permitted by the website or service) a combination lower case letters, uppercase letters, numbers, punctuation or other special characters to make up a password. The longer the password is, the long it will take for someone to guess or hack your password. This is an area where I am probably as guilty as everyone else is on this. I have a "favorite" password or two that I liked to use in the past. This kept me from having to have a way to keep track of all the passwords that I used on different websites. With more websites adding additional layers of security such as challenge questions, etc., I have had to start using a password manager app to help keep track of the different passwords, the challenge questions and answers used on a particular website, the recovery procedure if I am locked out of a website, what additional login procedures I have used for that site, etc.

Use a Unique Password

This one will be a bit of a pain but the time taken to do this will pay off in the long run. Look at it this way, if one of your logins on a particular site is compromised or hacked, you have taken a step to minimize the potential for additional website logins to be breached as well. Most password manager apps have the functionality of helping you generate a unique password. There are other apps that do this as well. A search of the app store for your phone or the internet for you laptop/desktop should show you some options to look at.

Different websites have different rules for how you can make your password strong. Some will let you use "special" characters such as * or +, others wont let you use those characters but will make sure that you don't use a password that can be looked up as a dictionary word. Some will not allow the same character or number to be used repeatedly. Remember, the more uniqueness you can have in a password makes it that much more difficult for someone to hack your account.

Enabling Two Factor Authentication

The only 2 factor authentication supported by LinkedIn is a SMS message giving you the 6 digit code to use to authenticate your login. While you can give your cell phone as the recipient of the messages, another option is to establish a Google Voice number as the receiver of the SMS messages. Doing so allows you to redirect where the SMS message goes so that you always have a backup way of getting the code.

After you have logged into your LinkedIn account, click on the Account and Settings menu option (typically has your picture on it). Click on Account at the bottom left hand area of the screen. Under Settings, Click on Manage Security Settings. Click on Turn on. You will be prompted to enter the phone number to send the SMS message to. Before this setting is allowed to take effect, you will need to enter the 6 digit code that should arrive shortly. I normally see the code within 10 to 15 seconds.

At this point, you have two factor authentication setup. Using the combination of a strong password and two factor authentication, you have taken all the steps possible at this point to protect your account from being hijacked.

Test Two Factor Authentication

- Logout of your LinkedIn account
- Log back in and test the two factor authentication
- Enter your username and password
- Follow the steps on the screen as indicated for the two factor authentication you have setup

Watch Video -
http://protectyourlogin.com/LI2FA

Twitter

Twitter is one of the more popular social media sites. It is also a prime candidate for having an account compromised. As with any social media or website, there are several things that can be done to make it a less than appealing target and encourage the would be attacker to move to another less prepared account.

Use a Strong Password

For those that aren't familiar with this term, a Strong Password is one that contains (where permitted by the website or service) a combination lower case letters, uppercase letters, numbers, punctuation or other special characters to make up a password. The longer the password is, the long it will take for someone to guess or hack your password. This is an area where I am probably as guilty as everyone else is on this. I have a "favorite" password or two that I liked to use in the past. This kept me from having to have a way to keep track of all the passwords that I used on different websites. With more websites adding additional layers of security such as challenge questions, etc., I have had to start using a password manager app to help keep track of the different passwords, the challenge questions and answers used on a particular website, the recovery procedure if I am locked out of a website, what additional login procedures I have used for that site, etc.

Use a Unique Password

This one will be a bit of a pain but the time taken to do this will pay off in the long run. Look at it this way, if one of your logins on a particular site is compromised or hacked, you have taken a step to minimize the potential for additional website logins to be breached as well. Most password manager apps have the functionality of helping you generate a unique password. There are other apps that do this as well. A search of the app store for your phone or the internet for you laptop/desktop should show you some options to look at.

Different websites have different rules for how you can make your password strong. Some will let you use "special" characters such as * or +, others wont let you use those characters but will make sure that you don't use a password that can be looked up as a dictionary word. Some will not allow the same character or number to be used repeatedly. Remember, the more uniqueness you can have in a password makes it that much more difficult for someone to hack your account.

Enabling Two Factor Authentication

The only 2 factor authentication supported by Twitter is an SMS message giving you the 6 digit code to use to authenticate your login. While you can give your cell phone as the recipient of the messages, another option is to establish a Google Voice number as the receiver of the SMS messages. Doing so allows you to redirect where the SMS message goes so that you always have a backup way of getting the code.

If you are going to create a Twitter account from scratch, this can be done automatically as a part of creating your account. If you skip that step or want to add it to an existing account, Click on Profile and settings icon (this can be your picture if you have one on the account). Go down the list of options and click on Settings.

If you didn't give your mobile number during the account creation process, you will need to enter that into the account settings first before you will be able to enable two factor authentication. Click on Mobile menu option. Enter your smart phone number into the indicated field on the screen. Click on the Continue button. You should get a text message with a 6 digit number momentarily. Enter that number in the screen asking for verification. You phone should now show as activated.

The next is to click on Security and Privacy menu option. Look for the Login Verification section. You have the choice of having a SMS message send as a part of the verification or have the authentication request sent to the twitter app on your smartphone. Even though you have just had to verify a 6 digit code when setting up the phone on the account, you will be asked to verify that you received a message. After that, you will be asked to enter your password after which the current configuration will be saved.

Add an additional step before a password reset can be requested

There is another option in this area that it would be a good idea to enable. Check the box beside Require Personal Information to reset my password. If you do request a password reset, you will be asked to enter the phone number that you have associated with this account before the password process can be done. Click on the Save Changes button to save these changes.

Test Two Factor Authentication

- Logout of your Twitter account
- Log back in and test the two factor authentication
- Enter your username and password
- Follow the steps on the screen as indicated for the two factor authentication you have setup

Watch Video -

http://protectyourlogin.com/TW2FA

Tumblr

Use a Strong Password

For those that aren't familiar with this term, a Strong Password is one that contains (where permitted by the website or service) a combination lower case letters, uppercase letters, numbers, punctuation or other special characters to make up a password. The longer the password is, the long it will take for someone to guess or hack your password. This is an area where I am probably as guilty as everyone else is on this. I have a "favorite" password or two that I liked to use in the past. This kept me from having to have a way to keep track of all the passwords that I used on different websites. With more websites adding additional layers of security such as challenge questions, etc., I have had to start using a password manager app to help keep track of the different passwords, the challenge questions and answers used on a particular website, the recovery procedure if I am locked out of a website, what additional login procedures I have used for that site, etc.

Use a Unique Password

This one will be a bit of a pain but the time taken to do this will pay off in the long run. Look at it this way, if one of your logins on a particular site is compromised or hacked, you have taken a step to minimize the potential for additional website logins to be breached as well. Most password manager apps have the functionality of helping you generate a unique password. There are other apps that do this as well. A search of the app store for your phone or the internet for you laptop/desktop should show you some options to look at.

Different websites have different rules for how you can make your password strong. Some will let you use "special" characters such as * or +, others wont let you use those characters but will make sure that you don't use a password that can be looked up as a dictionary word. Some will not allow the same character or number to be used repeatedly. Remember, the more uniqueness you can have in a password makes it that much more difficult for someone to hack your account.

Enabling Two Factor Authentication

- Click "Settings" under the Account menu (looks like the outline of a head and shoulders of a person) at the top of the Dashboard.
- In the Security section, enable "Two-factor authentication."
- Enter your phone number.
- Enter the password for your account
- You will receive a 6 digit code that you will enter to confirm that you have received the code
- Now decide whether you'd like to receive the code via text or through an authenticator app (we're into Google Authenticator). We recommend both in case you need to use one as a backup.
- Click the button beside Generate code via

Authenticator App.

- Take a screen shot of the QR code and file in a safe place noting the service and user id associated with it
- Follow the procedure for the authenticator app you are using to activate the QR code while it still valid

Application Passwords

There are times such as when you link your Tumblr account with something like IFTTT and using 2FA (Two Factor Authentication) won't be an option. This is where an application password will be useful. You will want to generate a different Application Password for each situation like this that you have. In that way, if one app does get compromised, you have the ability to block access from that app without having to redo your authentication setup.

Email about account activity

This is an option that I like to take advantage of when it is available. It serves as a check and balance on your existing authentication settings so that if you start getting emails about account activity, you know that it is something that you need to take a look at. You can always turn it off at a later date but if you aren't getting a flood of emails about logins, then it would hurt to leave it enabled.

Test Two Factor Authentication

- Logout of your Tumblr account
- Log back in and test the two factor authentication
- Enter your username and password
- Follow the steps on the screen as indicated for the two factor authentication you have setup

With Tumblr, you have the option of using the Two Factor Authentication option you initially setup before using the FIDO U2F key. If you don't have that key available, click Use mobile authenticator option to use that option instead.

Watch Video -

http://protectyourlogin.com/Tu2FA

Vimeo

Vimeo is one of those sites where you have an option on how to login.

Option 1

If you have a Vimeo account, then you can strengthen your existing password and follow the link to use the instructions later in this book "For Sites that don't support Two Factor".

Option 2

Use your Facebook account that has Two Factor Authentication enabled to login with. The first time you login with this option, you will have to answer a couple of extra questions about the rights you want Vimeo to have to access your Facebook account with and about it auto posting on your Facebook timeline.

Use a Strong Password

For those that aren't familiar with this term, a Strong Password is one that contains (where permitted by the website or service) a combination lower case letters, uppercase letters, numbers, punctuation or other special characters to make up a password. The longer the password is, the long it will take for someone to guess or hack your password. This is an area where I am probably as guilty as everyone else is on this. I have a "favorite" password or two that I liked to use in the past. This kept me from having to have a way to keep track of all the passwords that I used on different websites. With more websites adding additional layers of security such as challenge questions, etc., I have had to start using a password manager app to help keep track of the different passwords, the challenge questions and answers used on a particular website, the recovery procedure if I am locked out of a website, what additional login procedures I have used for that site, etc.

Use a Unique Password

This one will be a bit of a pain but the time taken to do this will pay off in the long run. Look at it this way, if one of your logins on a particular site is compromised or hacked, you have taken a step to minimize the potential for additional website logins to be breached as well. Most password manager apps have the functionality of helping you generate a unique password. There are other apps that do this as well. A search of the app store for your phone or the internet for you laptop/desktop should show you some options to look at.

Different websites have different rules for how you can make your password strong. Some will let you use "special" characters such as * or +, others wont let you use those characters but will make sure that you don't use a password that can be looked up as a dictionary word. Some will not allow the same character or number to be used repeatedly. Remember, the more uniqueness you can have in a password makes it that much more difficult for someone to hack your account.

Enabling Two Factor Authentication

Instagram is one of those services that doesn't support any of the two factor systems that services such as Facebook or Dropbox do for an additional layer of protection. There is an option. You can use the steps outlined in the chapter "For Sites that don't support Two Factor".

YouTube

Since this is a Google property. You can use your Google Credentials to login. If you don't have a Google account, you can setup an account by clicking on this link to Setup up Gmail.

Watch Video -

http://protectyourlogin.com/go2fa

For Sites that don't support Two Factor

I can't and won't take credit for this but wanted to make sure this made it into the book. Not all sites support Two Factor Authentication or sending you a SMS message with a code to enter before logging you in. There is what I call a Poor Man's way to get very close to that. The folks at Yubico have come up with an interesting way to just that. This involves using a static password stored in your Yubico Standard Key. The key itself is $25 so this is a way you can get very close to two factor authentication isn't currently offered on a site that you want to use it on.

http://protectyourlogin.com/Yust

You will need to download the Yubikey Personalization Tool (http://www.yubico.com/pt).

Download the version for the desktop/laptop operating system that you are using. Here is a link to the video that walks you through the process (https://www.yubico.com/products/services-software/personalization-tools/static-password/). With potentially having two separate partitions/ slots available on the Yubikey you are using, you can have two different long static passwords.

After you have your Yubikey programmed, you will go to the website you want to login to that website that doesn't support Two Factor Authentication, login with your username and password. Insert your Yubikey into the USB Port on your desktop/laptop, press the button on the Yubikey for several seconds and you should see the static password you programmed into the key inserted into the password field. Proceed with the rest of the login process and you should be good to go.

Adding an additional layer of protection

When using a network that I don't control or haven't configured, I like to take an additional step just to make sure that my communications are safe. Before you start telling me that you are using SSL or HTTPS and everything is encrypted and safe, I will say that just because you see the indications that the little lock is on the web browser window, someone may be still watching what is being sent. In the corporate world, I have used devices that insert themselves into the middle of this type of traffic. You may have heard the term "man-in-the-middle" which normally refers to a type of attack on the internet but aptly describes the situation I am talking about. This is a fairly common device that is present in a growing number of company and commercial data centers.

Another reason having an additional layer of protection on what you are sending over the internet is that more companies/hotels that offer "free" internet are doing so over a wireless connection that gives you absolutely no protection against preying eyes to see what you may be doing. To put it in perspective, anyone with a laptop and a copy of Wireshark or other packet capturing utility may be able to see some or all of what you are doing.

One thing that I do anytime that I am on an Internet connect that I don't control or maintain that I take additional step(s) to make sure that I am safe on the internet. One way of this is to use a

VPN client. This is a piece of software that builds an encrypted tunnel or connection between the device you are using and connecting to the point where the tunnel terminates somewhere out on the internet which would be well beyond where the would be hacker or thief will be able to see your traffic.

There are several service providers to choose from. My favorite provider to use is known as "TunnelBear". One thing that I like here is that I can use the service each month using upwards of 500MB of traffic at no charge. At any given time, there may be a special offer than you can take advantage of that will get you a temporary bump in the amount of traffic you can send/receive over the VPN. In the past, it has involved sending out a twitter message worded in a particular way and you get credit for the additional amount of traffic within minutes. As with anything I look at, I always like having a backup plan or provider that I can use. The other provider that I use is ZPN Connect. As with TunnelBear, you get a specific amount of traffic to use each month. They may also have a special offer that will earn you a temporary increase in the amount of traffic you can send over their internet connection.

http://protectyourlogin.com/tunnelbear

I like to have more than one VPN provider to use for several reasons. At any given time, there may be a problem with a particular provider to where you cant establish a connection to their service. One time it might be that they are either at capacity or experiencing problems. It could also be that the internet connection that you using may have some sort of block or system in place that doesn't allow for a connection to be established to a particular provider. That is where having at least one additional VPN service provider to use can help you get around those situations where someone is trying to block a particular VPN provider or the provider is having service related problems.

You will be references to other VPN Clients and services to consider using in the companion workbook that can be use in conjunction with this book. Please go to http://protectyourlogin.com to find out how to purchase the workbook.

You can also use a hardware firewall such as the ZyXEL MWR102 or the Ubiquity EdgeRouter Lite as two ways to protect your internet connection. There are software firewalls that you can consider as well. There is nothing to be said that you can't be running both a software based firewall and a hardware based firewall to make it even hardware for someone to break in..

Using a safe system to access the internet

There are times that you need to access one of the sites/services that you have protected with strong authentication services. The best of protection systems wont mean much if you are on a system where there is a keylogger or other software that may be trying to watch what you are doing and grab your login credentials. For most folks, that may not be a problem but if you need to be absolutely sure, there are a few option to consider.

ZeusGard

This is a novel approach to an age old problem of having a known safe platform to access the internet from. ZuesGard is a USB bootable secure browsing environment. If you happen to be using a computer with a network interface card that ZuesGard doesn't recognize, they offer the option of a USB dongle that offers wireless access to get network access.

ZuesGard offers several advantages over a DIY solution. You don't have to worry about a lack of a CD drive keeping you from booting up into a secure environment. You can put put it on a keychain or put in a small carry case usually meant for USB flash drives. At less than $25 US, it is a very affordable option to have in your travel kit.

http://protectyourlogin.com/zgusb

http://protectyourlogin.com/zueswifi

Kangaroo

This is something that might not normally appear on your radar. This is a completely self contained computer running Windows 10 that needs only a monitor/TV with an HDMI input and a keyboard to work. If you happen to be in an environment without Wifi, there are several USB ethernet adapters that will allow you to connect via an RJ-45 connector to the internet. Out of the box, this device comes with a built in fingerprint reader to further restrict you can use this device. You can add more memory to the device by plugging in a Micro SD memory card.

Kangaroo, from the folks at InFocus (a well known video projector company), can be a handle device to have nearby for those situations when you would like your regular desktop but don't want to lug around a laptop or other portable computing device. Currently this is available only via NewEgg or the Microsoft Store. For less than $100, it is worth considering to having in your bag of tricks.

http://www.kangaroo.cc/

Bootable CD

No discussion of environments would be complete without mentioning the Bootable CD. Available on more and more Linux distributions, computers are starting to ship without a CD drive. Normally this would be a show stopper but there are options where you can "burn" a copy of a bootable linux distributions on a USB Flash drive. The only thing you have to remember here is that older desktop/laptop hardware wont understand how to boot from a USB drive. Depending on how the device is configured, you may have to press F11 or other key to allow you to access a boot menu on that computer to give you a chance at selecting the USB Flash drive as the device you want to boot from.

While you may already have the tools to create your own bootable flash drive, there are two options that you want to look at if you don't already have something like this to use. Trinity Rescue Kit (trinityhome.org) and Rufus (rufus.akeo.ie) are tools to look at for this situation. While there are other tools available that can do this task, the names for these tools kept coming up when i was researching this part of the book. If you aren't on a newer change that comes with USB 3.0, this solution will still work with a USB v2 port, just a little slower.

There are a variety of linux distributions to choose from. The beauty of a bootable CD is that you can try a variety of these without going through the installation process. I have used Ubuntu (www.ubuntu.com) and SUSE (www.suse.com) over the years. There are a variety of Linux distributions that you can use. Don't worry about settling on just one. I usually have two or three in my bag of tricks. There are minimal distress (short for distributions) that install the bare minimum for those situations where you only want command line. There are Full Blown distros that give you a full graphical user interface to work with. As you look at Linux Distributions in more detail, there are "purpose built" distros specifically targeted at computer forensic investigations. Try different distros until you find one that you like and then learn what you can really do with it.

Your Journey starts now

Protecting your Online Presence is something that you need to start as soon as you can. This will be a process that you will need revisit on a periodic basis even if nothing else changes. Depending on the number of websites that you may be using, periodically checking these websites when they do any type of upgrades or make any changes to their privacy policies also needs to be done. This might seem like it will take a lot of time but would you rather spend a little bit of time now preventing a problem or more time later dealing with the aftermath of a problem.

I plan on releasing updates to this book. Please check my website http://www.ronaldnutter.com or http://protectyourlogin.com for information on updates as they are released or notifications on when an updated version of this eBook will be available via the Amazon Kindle website. Also, be sure to get on my email list for this book using the link at the front of the book for automatic notification of updates.

Please check out my other eBooks and add them to your library today !

This book is targeted at the new VMware Admin or someone who has been out of the field for a while and needs to come up to speed fast. Think of this as an electronic notebook with things you won't learn in a VMware class. If you are looking for a job in this area, some of what you find here might just make you stand out from the others interviewing for the position.

http://www.ronnutter.com/vmna1

If you have dealt with any type of technology, you will know there are times when using the GUI won't work. This book gets you familiar with the different ways of accessing either a VMware host, vCenter server or VDP appliance. Knowing how to navigate around the Command Line Interface in the VMware product family is a handy skill to have.

http://www.ronnutter.com/vmna2

At some point in your VMware career, you will need to do at least on upgrade. While no two upgrades may ever be the same, having a rough blueprint will make the process a little easier. This is another area where doing a dry run in your lab will help give you an idea of what to expect in the real world. Running into problems in the lab is easier to deal with than in a production environment.

http://www.ronnutter.com/vmna3

In volume 4 in the VMware - A Guide for New Admins series, I put together a timeline for those with little to no background in VMware to help them get started on the path on learning VMware and a new job skill. This book will be updated as I release additional books in the series to continue the buildout of the path for increasing your VMware skills.

http://www.ronnutter.com/vmna4

This book is targeted at the new VMware Admin or someone who has been out of the field for a while and needs to come up to speed fast. Volume 5 deals specifically the different things you need to think about in backup up your growing VM farm. This is another area where you will need to be well versed in.

http://www.ronnutter.com/vmna5

You can't always depend on a cell phone during an emergency or disaster. This book covers a variety of options such as Satellite, Amateur Radio, MURs, GMRS and several other options. One thing you need to read about is how to use your cell phone to establish your own private texting network !

http://www.ronnutter.com/ppv1c

This book covers steps you will need to take to protect your personal information during a disaster. This is a prime time for identity thieves to strike. This will also be the same time when you will have the least time and resources to deal with the situation.

http://www.ronnutter.com/ppv2p

Depending on the length of a disaster or emergency situation might mean that the grocery store(s) near you will be low or out of food. Know how to do some basic food preparation steps or even growing some of your own foot before a problem occurs will help you be that much more self sufficient for the duration of the disaster/ emergency and as a matter of general principle.

http://www.ronnutter.com/ppv3f

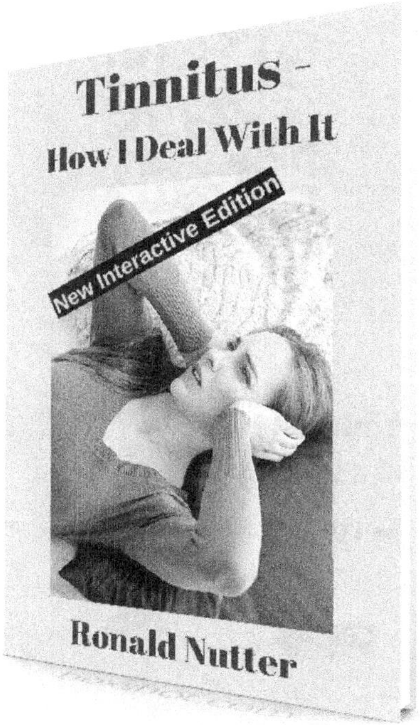

It is estimated that somewhere between 1 and 5 to upwards of 50 million people suffer with Tinnitus, otherwise known as ringing in the ears. In the past, there wasn't much that could be done to treat Tinnitus as there currently is no cure for the problem. That all changed in early 2014. This book is a diary of my journey since being diagnosed with Tinnitus and what I have learned along the way.

http://www.ronnutter.com/tims

Android isn't just an operating system for your phone or tablet. There are a variety of apps that can turn your device into a ad-hoc network testing tool so you don't have to carry a war chest of testing equipment and tools with you. With the tools listed here, you can be ready to troubleshoot at a moments notice !

http://www.ronnutter.com/asne

As the stories continue to show up in the press about websites being broken into or accounts being hacked, you can have a variety of options to look at to protect yourself. There are a variety of options that until recently were available only to medium to large commercial clients. This book walks you through step by step on ways that you can make your accounts just hard enough to try to break into that the hackers will look elsewhere for easier less protected targets.

http://www.ronnutter.com/thac

Made in the USA
Monee, IL
04 November 2026

67376428R00095